Christina Mayer

Feeding aggression in dogs (Canis familiaris) and wolves (Canis lupus)

AF138586

Christina Mayer

Feeding aggression in dogs (Canis familiaris) and wolves (Canis lupus)

A comparison between dog packs and a wolf pack of different age classes

Natural Sciences Series

Impressum / Imprint
Bibliografische Information der Deutschen Nationalbibliothek: Die Deutsche Nationalbibliothek verzeichnet diese Publikation in der Deutschen Nationalbibliografie; detaillierte bibliografische Daten sind im Internet über http://dnb.d-nb.de abrufbar.
Alle in diesem Buch genannten Marken und Produktnamen unterliegen warenzeichen-, marken- oder patentrechtlichem Schutz bzw. sind Warenzeichen oder eingetragene Warenzeichen der jeweiligen Inhaber. Die Wiedergabe von Marken, Produktnamen, Gebrauchsnamen, Handelsnamen, Warenbezeichnungen u.s.w. in diesem Werk berechtigt auch ohne besondere Kennzeichnung nicht zu der Annahme, dass solche Namen im Sinne der Warenzeichen- und Markenschutzgesetzgebung als frei zu betrachten wären und daher von jedermann benutzt werden dürften.

Bibliographic information published by the Deutsche Nationalbibliothek: The Deutsche Nationalbibliothek lists this publication in the Deutsche Nationalbibliografie; detailed bibliographic data are available in the Internet at http://dnb.d-nb.de.
Any brand names and product names mentioned in this book are subject to trademark, brand or patent protection and are trademarks or registered trademarks of their respective holders. The use of brand names, product names, common names, trade names, product descriptions etc. even without a particular marking in this works is in no way to be construed to mean that such names may be regarded as unrestricted in respect of trademark and brand protection legislation and could thus be used by anyone.

Coverbild / Cover image: www.ingimage.com

Verlag / Publisher:
AV Akademikerverlag
ist ein Imprint der / is a trademark of
OmniScriptum GmbH & Co. KG
Heinrich-Böcking-Str. 6-8, 66121 Saarbrücken, Deutschland / Germany
Email: info@akademikerverlag.de

Herstellung: siehe letzte Seite /
Printed at: see last page
ISBN: 978-3-639-49689-5

Table of Contents

List of Tables

List of Figures

1 Introduction

1.1 Aggression: its forms and functions

Aggression is a basic motivational and behaviour complex in all animals that has several different forms and plays an important role in the survival and reproduction of animals. The degree of aggression and the outcome of it depend on the urgency and motivation of the animal (Lorenz 1964). Additionally there are different forms of aggression which depends on the situation. In most of the cases it is proximity-induced, for example in the case of feeding, but it could also be an expression of frustration, pain or fear (Hinde 1970, Lockwood 1995). Alongside submissive and fleeing behaviour, aggression is a form of agonistic behaviour: it may occur when two or more animals interact in a conflict situation (Lorenz 1964, Pal et al. 1998, Langbein & Puppe 2004). Though different species are routinely described by different levels of aggression (Thierry 2000), it is important to differentiate between intergroup and within-group aggression because their levels vary independently across species (e.g. Miklósi 2007 pp. 81). In intergroup conflicts two or more different groups of animals show aggressive behaviour, mostly when competing for territories or other resources. The conflicting groups typically do not provide benefits for each other, therefore only trying to avoid getting injured limits how serious these fights become. Thus, in many species, including wolves and chimpanzees *(Pan troglodytes)*, fights between groups are lethal (Mech & Biotani 2003; Wilson & Wrangham 2003; Mitani et al. 2010).

Within-group aggression also arises from competition for limited resources, like food and mating partners in this case, however, between members of the same group (Brown 1964, Archer 1988, Saito et al. 1998, Feddersen-Petersen 2004). In this case it is usually assumed that increased defence against predators and other benefits of group living (e.g. cooperative hunting or breeding) compensate for competition within a group (van Schaik 1989). Consequently, the aim of within-group aggression usually is to solve the current conflict without killing a group member or even causing damage to its physical condition or to the relationship the aggressor has with its opponent (Hinde 1970, Feddersen-Petersen 2004). As such, within-group aggression can be ritualized to a high degree to ensure that the animals in conflict do not hurt each other severely. This has been described in wolves for instance, where aggression can range from staring intently at another animal through barking or growling to chasing, pushing away and finally snapping and only ultimately to real fighting with actual physical contact that may in extreme cases cause an injury or death (Mech 1970; Mech & Boitani 2003). Displays of dominance and submission as well as aggressive threats and fleeing are important devices of conflict management without incurring high costs on either competitor (Preuschoft & van Schaik 2000).

Going beyond these general principles, there is a great variability across species regarding the severity and symmetry of their aggressive interactions. This cannot be explained exclusively by evolutionarily relatedness of the species because even closely related species can strongly differ. In macaques for example, Thierry (2000) described a four-grade scale were he arranged macaque species according to their aggression, tolerance, conciliatory behaviour, dominance gradient and kin-bias. Mainly rhesus *(Macaca mulatta)* and Japanese macaques *(Macaca fuscata)* belong to the first grade, because of their unidirectional aggression with high and severe biting rates. Dominant animals are highly aggressive and subordinates generally flee or submit when attacked. The dominance gradient in these species is the steepest and the hierarchy is rigid with less contact between animals whose status is far apart. Thierry rated Tonkean *(Macaca tonkeana)*, moor *(Macaca maurus)* and crested macaques *(Macaca nigra)* in the fourth grade. The intensity of aggression and the biting rate are low, and most aggressive interactions are bidirectional, meaning that the victim of aggression protests or counter-attacks. In these species, dominance ranks are also stable but less steep. The animals have frequent contact to each other no matter which status they have. The other macaque species are rated between these two extremes. It can be said that the asymmetry of contests and the dominance gradient decrease from the first to the forth grade and social tolerance increases.

It has been proposed that the different socio-ecological conditions (e.g. food distribution, strength of between-group competition, distribution and density of predators) are responsible for the behavioural variation macaques have evolved (van Schaik 1989). For example the distribution of food triggers different behaviours. When the food is clumped there is more aggression found in the animals and when the food is more distributed the animals show more tolerant behaviours and feed more relaxed next to each other because in this situation it is more difficult to monopolize the food (van Schaik & van Noordwijk 1988). Others, however, argue that phylogenetic relatedness does play an important role (Thierry 2007). As Thierry (2007) pointed out, macaques in the same grade, as described above, are likely to be closer related than those from different grades.

Independently from the evolutionary origin of this behavioural variation, it seems that a range of social behaviours co-vary across species and form interconnected sets of traits. It has been proposed that these behaviours are linked due to social epigenetic processes influencing all of them (Sterck et al. 1997, Thierry 2004). For example a reduced serotonin activity was found in connection with a higher amount of aggression (Kaplan et al. 1995). Modelling studies, however, demonstrated that even in simply structured groups of virtual entities varying only the dominance gradient influences the spatial structure and other social characteristics of the groups, questioning the role of underlying biological mechanisms (Hemelrijk 1999). Independently from their proximate cause, it is widely accepted that different social behaviours, ranging from dominance, aggression, through social attentiveness, vigilance to feeding behaviour are interconnected and co-vary, which has been demonstrated not only in mammals but also in birds (Kotrschal et al. 1993). Even more, it has been proposed that if only some of these behavioural characteristics of a species with known socio-ecology and/or phylogenetic relatedness are known, based on such regularities predictions can be made regarding other elements of these behavioural complexes (Thierry 2000).

Based on this argument, it becomes more understandable that species characterizations are readily made after observing animals in one or a few contexts. Within-group aggression and tolerance are often described during group feeding, since competition for food is probably the most frequent and universal source of conflict in animal groups. Indeed, it has been shown that tolerant groups co-feed in a calmer way in contrast to animals living in a despotic system where usually one or more animals monopolize the food (e.g. long-tailed macaques *(Macaca fascicularis)*: van Schaik & van Noordwijk 1988, van Schaik 1989). It has also been suggested that animals that hunt in groups are more tolerant because they need to cooperate during the hunt in order to get food (e.g. wolves *(Canis lupus)*: Zimen 1990, Mech & Boitani 2003; African hunting dogs *(Lycaon pictus)*: Kühme 1965). Also some animal species that raise their offspring cooperatively, like common marmosets *(Callithrix jacchus)*, are more tolerant in food related situations (Burkart et al. 2007) (though in other species this link seems to be missing, e.g. cottontop tamarins *(Saguinus oedipus)* are also cooperative breeders but they do not donate food to those animals from which they got food before. Additionally they prefer to eat the food themselves or even do not pull the rope, where the food dish was attached, when their partner signalled them that they would need help from them (Cronin et al. 2009). Whereas on the contrary, capuchin monkeys *(Cebus apella)* that are not cooperative breeders do show high tolerance during feeding, including food sharing (de Waal 1997)). The general belief that co-feeding is a useful way to describe the aggressiveness or tolerance of the participants is reflected by the fact that on the field of animal behaviour a tolerance test typically means confronting two animals over a limited food resource (Hare et al. 2007).

1.2 Tolerance and its implications for cooperation

Similarly to the macaques, also chimpanzees and bonobos *(Pan paniscus)*, another pair of closely related primate species, strongly differ in regard to tolerance. Both species live in large communities with a fission-fusion characteristic where generally the males stay in their birth group and the females disperse and join neighbouring groups, both species are promiscuous with strong bonds between family members but also with non related animals. Still, the behaviour of chimpanzees and bonobos in competitive situations is very different (Hare 2009, Furuichi 2011). Chimpanzees attack and bite their group members more severely than bonobos that express aggression at a low intensity and bite rarely (Wrangham et al. 2006; Furuichi 2011). Bonobos are more tolerant than chimpanzees in feeding situations (Hare et al. 2007) and they share food voluntarily (Hare & Kwetuenda 2010). Importantly, Hare and his colleagues (2007) demonstrated that the level of tolerance and their success in a cooperative task co-varies in chimpanzees and bonobos. They had a closer look on the cooperative abilities of chimpanzees and bonobos in a co-feeding experiment. First, dyads of the same species were co-feed in three different conditions where either both dishes were baited or one dish was baited with sliced food and the other one was empty or one dish was baited with two pieces of food and the other was empty. As long as both dishes contained food, both the chimpanzee and bonobo dyads ate more or less peacefully together. However, when there was only one baited dish Bonobos were more tolerant than chimpanzees. Bonobos showed more socio-positive and play behaviour and less aggression than chimpanzees. Afterwards, the dyads were confronted with a food platform that two animals needed to pull simultaneously with the help of a rope to pull it close to the cage and to get the food placed on it. When the food was highly shareable, namely slices of food at two dishes, both species cooperated at more or less the same level but bonobos performed much better when there was just one dish baited with two pieces. In these conditions where the food was monopolizable, bonobos showed less aggression and more often socio-sexual and play behaviour. Chimpanzees seemed to avoid each other and thus, the dyads did not come close enough to the platform together so that they could have successfully solved the task. Hare et al. (2007) proposed that the flexibility of chimpanzee cooperation was constrained by the lack of tolerance between the two partners. When these constraints were removed the chimpanzees could show similar cooperation forms like bonobos (see also Melis et al. 2006). Additionally, chimpanzees cooperated most successfully with conspecifics with whom they had high tolerant relationships (shared food also in other situations) and whom they were related to (like parent-child dyads). Consequently, cooperation can be constrained by non-cognitive processes, such as motivation and tolerance to the partner.

A similar connection between species-specific tolerance and cooperation has been shown in macaques as well (Hare 2009). Tonkean macaques *(Macaca tonkeana)* with relaxed hierarchies and egalitarian and tolerant social organization are more successful when cooperation is needed. In contrast, rhesus macaques *(Macaca mulatta)* that have strict dominance hierarchies and despotic and less tolerant social groups fail to cooperate because in cooperative situations the dominant cannot inhibit their aggressive behaviours.

Based on these findings, Hare and his colleagues (2005) proposed that selection for increased tolerance (or tamer temperament or lower emotional reactivity in other words) might have been a prerequisite of the evolution of advanced cooperative behaviours. This theory, named as the emotional-reactivity hypothesis, suggested that social problem-solving might have evolved as a by-product through selection on emotional systems, such as those controlling the expression of fear and aggression (Hare et al. 2005). As such, the hypothesis turned the focus of research exploring the evolutionary origins of advanced cooperation from cognitive abilities to affective mechanisms. The theory has been extended on humans as well, claiming that similar processes might have been important also during the evolution of the unique cooperative abilities of humans in contrast to one of our closest living relatives, the chimpanzees (Hare & Tomasello 2005). Hare and his colleagues (2012) suggested that in various species that went through selection for increased tolerance, a more or less interconnected set of traits evolved that form the so-called domestication syndrome. In line with the above-described model of co-varying social behaviours of macaques, their proposal predicts that species selected for increased tolerance show similar morphological, developmental, behavioural and cognitive changes, including increased cooperativeness. They list humans, bonobos, urban and domestic animals as potential candidates that went through such evolution.

1.3 Tolerance and cooperation in dogs and wolves

As argued in the previous section, the emotional reactivity hypothesis is a parsimonious way to explain the evolutionary origins of the exceptional cooperative skills of humans. Importantly, Hare and Tomasello (2005) used the domestic dog as another case (along human-chimpanzee differences) to confirm the emotional reactivity hypothesis.

According to previous archaeological findings (Davis & Valler 1978, Clutton-Brock 1995) and recent genetic studies (Savolainen et al. 2002, Pang et al. 2009, Gray & Wayne 2010) the domestication of the dogs began at least 14,000-16,300 years ago. However, in 2009, some geneticists (Germonpré et al. 2009) analysed canid sculls from Belgium and they found out that they were at least 30,000 years old, which matched also with the findings by vonHoldt et al. in 2010. This debate to date remains unresolved. Behavioural as well as genetic studies agree, however, that the closest wild-living relative of dogs is the wolf (Clutton-Brock 1995, Pang et al. 2009, Gray et al. 2010). Very relevant for our topic, some researchers suggested that at the beginning of domestication the wolf kind of selected itself for tameness: less fearful wolves stayed close to human settlements, lived on the garbage humans produced and mated with each other thereby producing a population of wolves that showed decreased fear and aggression toward people (Morey 1994; Coppinger & Coppinger 2001). This scenario can provide a good basis for the emotional reactivity hypothesis that suggests that dogs are less fearful and aggressive (that is more tolerant) and also more cooperative than wolves are (Hare et al. 2012). Regarding cooperativeness, however, the emotional reactivity hypothesis focuses exclusively on how wolves and dogs can cooperate with humans.

For instance, dogs perform better in reading humans' cooperative-communicative cues than humans' closest relatives, the apes: they readily follow pointing and gazing when a human is indicating in which of two containers the subject can find food (Bräuer et al. 2006, Soproni et al. 2001). Even more importantly, it has been shown that wolves, even if human-raised, need longer time to develop these skills in contrast to dog pups that readily use human cues, especially pointing, in a food-searching task

(Hare et al. 2002, Miklosi et al. 2003, Gácsi et al. 2009). Thus, it has been suggested that dogs and humans went through convergent evolution (Miklosi et al. 2004; Hare & Tomasello 2005).

Further supporting the emotional reactivity hypothesis, comparisons with experimentally selected foxes hint into the direction that these abilities of dogs evolved as a by-product of selection for tame behaviour (Hare et al. in 2005). Belyaev (1979) selected a group of foxes against fear and aggression toward humans: bred always those animals that readily approached the hands of humans standing in front of their cages. In comparison with a control group of foxes that was bred randomly regarding their behaviour towards human, the experimental group was more skilled in using human gestures in finding food (Hare et al. 2005). They performed similarly to pet dogs. Additionally they showed enhanced social tolerance and cooperative abilities compared to the control foxes. Further on, the foxes selected for tameness not only showed dog-like behaviours but also a lower level of cortisol and adrenocorticotrophic hormone (ACTH) in the blood plasma, and their adrenal response to stress was reduced in comparison to the control group of foxes (Trut 1999, Gulevich et al. 2004). The down-regulation of the hypothalamic-pituitary-adrenal axis that is responsible for stress reactions, fight or flight (Toates 1995, Tsigos & Chrousos 2002) can easily be linked to selection for less fearful behaviour (Oskina 1996, Künzl & Sachser 1999). Other physiological changes were also recorded, such as different gene-expressions and altered responsiveness to serotonin, noradrenaline and dopamine in specific brain regions that are involved in the regulation of emotional-defensive responses (Saetre et al. 2004, Popova et al 1991, Trut et al 2000).

Based on the covariation of reduced fear and aggression, increased tolerance and increased cooperative-communicative skills in the experimental foxes, Hare and Tomasello (2005) proposed that a similar selection for a tamer temperament (increased tolerance) in the domestic dog is sufficient to explain its increased cooperativeness.

This hypothesis, however, is based on a very limited set of data, and provides no information regarding dogs' and wolves' skills for cooperation with conspecifics. Observations on free-living wolf and dog packs offer a very different picture compared to interactions with humans. While free-living dogs and wolves seem to form a similar social hierarchy, their groups rely on cooperation to a very different extent. In the wolf (*Canis lupus*) (Schenkel 1947; Zimen 1990) as well as in the dog (*Canis familiaris*) (Pal et al. 1998, Bonanni et al. 2010, Cafazzo et al. 2010) we can find linear dominance hierarchies, divided into a male and a female line. Wolves are cooperative breeders, meaning that the whole pack contributes to the defence, feeding and raising of the pups. In the first 3 weeks of the pups, all pack members help feeding the mother that stays in the den located in the middle of the territory, and later on they regurgitate food for the begging pups (Mech 1970). In contrast, in feral dogs the mothers usually give birth alone on the periphery of the territory (Kleiman & Malcolm 1981, Daniels & Bekoff 1989, Boitani et al. 1995). The father and the pack mostly do not contribute to the raising of the young (Miklósi 2007), except for feral dogs in West India where fathers were observed feeding the puppies and protecting the den (Pal 2003, 2004, Pal et al. 1998). Cooperation during hunting is a main part of the group living of wolves (Mech 1970). In feral dogs, however, the ability to hunt large prey in groups seems to be less effective (Boitani & Ciucci 1995), or is simply unnecessary because they have a lot other and less exhaustive possibilities to get food (Berman & Dunbar 1983). In sum, when it comes to comparing within-group cooperation with conspecifics it is questionable whether the presumption of the emotional reactivity hypothesis that dogs are more cooperative than wolves is valid or not.

Similarly contradicting pieces of information are available regarding the aggressiveness and tolerance of dogs and wolves. Hare and colleagues (2012) reviewed that female wolves readily attack the pups of subordinate females or even kill them (McLeod 1990; Sands & Creel 2004). This behaviour was never reported in feral dogs (Boitani & Ciucci 1995). Also in between group interactions killings of wolves are frequently observed (Mech 1994, Mech et al. 1998) whereas dogs prefer to bark at each other until one group backs out (Boitani et al. 1995). As argued in the beginning of the introduction, however, between-group aggression is likely to be less relevant for within-group strategies of conflict resolution. Very few studies exist that compared interactions in dog and wolf packs living under comparable conditions. Feddersen-Petersen (1991) reported fierce within-group aggression, which she observed in her poodle group, and concluded that it resulted from a lack of a "fine-tiered social hierarchy". In line with her observations, some argue that a consequence of domestication is that dogs show more overt aggression against conspecifics compared to wolves (Frank & Frank 1982, Feddersen-Petersen 1991, Goodwin et al. 1997).

Dogs are different from wolves in many ways. Wolves live in packs of two to 36 individuals, but the average number of individuals in one pack is about six (Schenkel 1947, Mech 1970, Zimen 1990). As described above, for wolves pack-living is likely to increase their fitness: helps to hunt enough food to survive, to rear pups to bring their genes into the next generation or to defend themselves against competitive packs (Mech 1970, Mech and Boitani 2003). In contrast, for most dogs humans are their partners. Dogs do not need to hunt or to make other important decisions, since their food is delivered by humans and humans decide what is happening, where they live or go and if and with whom they should mate. Their survival is dependent on humans instead of conspecifics. Therefore, it seems that for dogs it is not necessary to form social bonds with other dogs – instead they form bonds with humans (Boitani et al. 1995) and they pay attention to the subtle signals of humans rather than on conspecifics (Zimen 1992). Similarly, aggressive interactions between dogs might also have been influenced by domestication, since humans typically intervene during fights and provide the dogs care afterwards. This decreases the negative consequences of initiating fights in dogs, and, ultimately, might have lead to more severe escalation of aggression in the domestic dog in comparison to the wolf (Goodwin et al. 1997). Others proposed that also the morphological changes especially of the face, ears and general body posture which most dog breeds gathered during domestication may make the signals of dogs less clear and thus, communication more difficult (Feddersen-Petersen 1991, Clutton-Brock 1995, Leaver & Reimchen 2008, Kerswell et al. 2009).

1.4 Research Questions and Hypotheses

In sum, sharply contradicting hypotheses regarding dog-wolf differences and similarities in their aggressiveness exist, and relevant behavioural data is scarce. As described in the introduction, I aim at investigating the following two contradicting hypotheses.

"Relaxed selection" hypothesis: Since, due to domestication, dogs do not need to rely on their interactions and cooperation with conspecifics any more, the selection pressure on fine-tuning their communication and inhibiting their aggression has relaxed. This hypothesis predicts that dogs show more frequent and especially stronger aggressive behaviour than wolves do (Frank 1980, Zimen 1992, Boitani et al. 1995).

"Emotional reactivity" hypothesis (Hare and Tomasello 2005): Through the evolution from the wolf to the dog, first the wolves selected themselves against shyness which is called the "self-domestication hypothesis" (reviewed in Hare et al. 2012) and these already tamer wolves were actively selected against fear and aggression through the humans and therefore dogs should be more tolerant than wolves, or in other words, wolves should show more aggression than dogs.

In this study I will investigate if there is a difference between the feeding behaviour and aggression of dogs and wolves living in packs and raised similarly. My aim is to provide the first extensive analysis on aggression and tolerance in dogs and wolves in a co-feeding situation in a way that is comparable to wider species comparisons. Accordingly, I will analyse the following measures during group feeding: the duration of aggression, to see if there is a difference in the time the animals spent with the weak, intermediate and strong aggression and the frequency of aggression, to see how often they use aggression in the feeding context and to include behaviours that are not measured in time but just in the quantity they occurred like for example bite, snap or growl. First of all I will look at the differences of dogs and wolves, but also if there is a difference between the dominant animals and the subordinates and as well if there is an effect of the sex and the age.

Additionally I will test the animals in 4 different conditions to see if their behaviour changes from monopolizabel food, like one and two pieces, to more distributed food, like one piece per animal and several small pieces.

2 Material & Methods

2.1 Subjects

I observed 2 packs of mixed-breed dogs (*Canis familiaris*) and 1 pack of timber wolves (*Canis lupus occidentalis*) composed of animals of two age classes. During the video recording (from April to August 2011 for the dogs and from October to December 2009 for the wolves), the adult animals were about 1.5 years old, and the young ones were about half a year old. Dog pack 1 consisted of 2 adults and 4 pups (Table 1) and dog pack 2 consisted of one adult and one pup (Table 2). The wolf pack consisted of 3 adult animals and 6 pups (Table 3). All animals were hand raised in peer groups from their age of 1 to 2 weeks on by a group of 3 to 9 people. They were bottle-fed and later hand-fed by people and for 3 months at least one hand-raiser was present in the puppy enclosure for 20 to 24 hours daily. The pups were raised in an enclosure next to the adult animals, thus they could see and hear each other from the beginning on and they were brought into direct contact regularly for introductory sessions of 30 to 60 min each. In months 4 and 5 human contact was gradually reduced till the point when the young animals were introduced into the pack of the older animals. From then on the 2 generations lived together in the same enclosure. However, all animals went on having daily face-to-face interactions with humans in frame of training sessions and cognitive and behavioural tests.

Table 1: The birth date, relatedness and sex of the members of dog pack 1.

Subject	Birth date	Litter	Sex
Rafiki	29.11.2009	1	male
Maisha	18.12.2009	2	male
Asali	13.09.2010	3	male
Binti	13.09.2010	3	female
Bashira	13.09.2010	4	female
Hakima	13.09.2010	4	male

Table 2: The birth date, relatedness and sex of the members of dog pack 2.

Subject	Birth date	Litter	Sex
Kilio	18.12.2009	2	male
Meru	01.10.2010	5	male

The dogs were fed daily with dry food and occasionally with small pieces of meat or whole carcasses (rabbit or goose). The wolves were fed every second or third day in winter and once a week in summer with pieces of meat or whole carcasses (deer, rabbit, chicken). Water was always available in all enclosures.

Table 3: The birth date, relatedness and sex of the members of the wolf pack.

Subject	Birth date	Litter	Sex
Kaspar	04.05.2008	1	male
Aragorn	04.05.2008	2	male
Shima	04.05.2008	2	female
Nanuk	28.04.2009	3	male
Geronimo	02.05.2009	4	male
Yukon	02.05.2009	4	female
Tatonga	21.04.2009	5	female
Apache	19.05.2009	6	male
Cherokee	19.05.2009	6	male

2.2 Data collection

Eighteen feeding sessions were recorded in the wolf pack, and 20 feedings in each of the 2 dog packs. Depending on the size and number of food pieces offered, each feeding session belonged to one of 4 different feeding conditions: 1) a single big piece of meat or carcass for the whole pack, 2) two big pieces of meat, 3) one piece of food for each member of the pack and 4) several small pieces of food (Table 4). The order of the different feeding conditions was randomly chosen. In all conditions the food was thrown over the fence on a flat place in the home enclosure and was distributed in the "one piece each" and "several pieces" condition, on a place of about 4 x 3 m.

Table 4: Amount of feedings per condition in all packs.

Pack	1 big piece	2 big pieces	One piece each	Several pieces
Wolves	4	3	3	8
Dogs 1	3	3	4	10
Dogs 2	4	3	3	10

Two experimenters filmed the animals from outside the enclosures with video cameras (Sony Handycam DCR-SR35). They were positioned in front of the place where the food was thrown in and on the other side of the enclosure. One experimenter focused on one animal at a time to record her or his behaviour for 10 minutes (focal animal sampling) and then switched to another animal to record the behaviour of them for another 10 minutes and so on until all animals were recorded or the feeding session ended. The order of recording the different animals was predetermined and randomized across feeding sessions. The second experimenter recorded the whole pack in order to catch all aggressive interactions. Records ended when all food was gone or after 1.5 hours after the food was provided.

2.3 Behavioural coding

The dog and wolf videos were analyzed with the Noldus Observer XT 10.

The feeding time started when the food was thrown over the fence and ended when there was no food left or after latest 1.5 hours. In this case either the food piece was too big for them to finish, like a deer, or when the animals stopped eating and didn't continue until the 1.5 hours were over. Additionally, I was interested how much time each animal spent at a certain distance to the food and, consequently to an eating animal.

I used the ethogram developed by Matthey-Doret (2010) (based on Goodmann & Klinghammer 1990, Koler-Matznick et al 2005, WSC ethogram 2008 and Möslinger 2008 (Table 5, Appendix)) to record the aggressive interactions of the animals. Aggressive behaviour was categorized into 3 classes. "Weak aggression" included all aggressive behaviours with no physical contact between the interacting animals. In "intermediate aggression" physical contact was involved but the intensity was low and did not cause any damage. Finally, in "strong aggression" all behaviour and physical contact between two or more opponents was summed up which had the potential of hurting each other (for example attack or bite) or challenged the opponent physically with a high intensity for an extended time (for example jaw spar) (Matthey-Doret 2010) (Table 5, Appendix). The total duration and the total number of all observation were extracted from the coded behaviours for the analysis.

In order to establish the dominance hierarchy in each pack, I used the dominant and submissive behaviours shown below (Table 5).

Table 5: Ethogram. Description of the behavioural categories.

Levels	Behaviours	Descriptions
	Feeding time	time the animal spend with eating
	Distance to food	range of meter between the animal and the food (next: <1 m, close: 1-3 m, distant: >3 m)
	start-stop	the coded observation started at the time the food was thrown over the fence and ended when all food was gone or after 1.5 hours after the food was provided
Aggressive Behaviours		
Weak aggression		
	Ambush	lying in a sphinx posture, staring intently at another wolf or at prey
	Bark	a short explosive outburst and coarse voice, sound like wuff
	Besnuffle	to sniff in an exaggerated way to another subject, tends to precede bites
	Chase	running in pursuit of another
	Dominant approach	to go forward within 2 m to another individual with the tail perpendicularly or above the plane of the back and the ears erects and pointed forward and head held high
	Growl	a throaty rumbling vocalization, usually

	low in pitch
Guard	to stay by something and to drive others away from it
Stalk	to pursue another individual or prey by means of stealthy approach; the head level is lower than the top of the back with the ears directed forward

Intermediate aggression

Displace	aggressor causes opponent to move away from a resource or goal
Grab	to bite another individual, and hold on firmly
Inhibited bite	a bite without sufficient pressure to wound a conspecific
Muzzle bite	grabbing the muzzle of another animal with the jaw without hard pressure
Pin	to grab another individual forcing it to the ground and holding it there
Pull	to grab another individual and draw it along, without the pulled individual being recumbent
Push away	to use feet and legs both defensively while engage in a ritualized aggression
Rebuff	rejecting a suitor and driving him or her away
Ride up	to mount another one from behind or from side
Snap	a rapid bite that has a little contact with its object; as the canid's jaws close, the teeth make an audible sound
Tug of War	two individuals holding of different parts of an object and tugging vigorously against each other

Strong aggression

| Attack | a running or jumping approach toward another one, often bites at the neck or muzzle forcing it on the ground and holding it there |

Bite	to close jaws and teeth on someone which can cause a wound to a conspecific
Fight	high intensity, aggressive, often damaging encounters
Hipslam	the individual pivots on its forepaws and slams into its target with its hindquarters
Jaw spar	two canids "fencing" with open jaws; as they block each other's feints, neither actually closes its jaws
Knock down	striking another individual with a sharp blow, usually with the chest and shoulders
Mob	chasing, jaw sparring, biting, and/or wrestling or pinning by two or more individuals orienting to a third

Submissive Behaviours

Active submission	the individual has its tail tucked between his hind legs, sometimes wags it while he is in a crouched position and may attempt to paw and lick the side of aggressor's muzzle and mostly pees
Flee	to walk or run with tail tucked and body ducked away from other individuals
Food beg	to lick, nibble, pull or paw at another individual's muzzle and lips
Crouch	to lower the head, bent the legs, the back often arched and the tail between the legs; the animal looks hunched and smaller
Passive submission	to lie on the back, demonstrate the stomach and the tail is between the legs; the ears are directed backwards and close to the head and raises a hind leg for inguinal presentation

Dominant Behaviours

Dominant approach	to go forward within 2 m to another

	individual with the tail perpendicularly or above the plane of the back and the ears erects and pointed forward and head held high
Genital sniffing	to sniff the genital parts of another individual when this one is lying on the back in the passive submission position
Mark	to urinate with the hind leg lifted up in the air mostly near or on bushes or on a tree
Muzzle bite	grabbing the muzzle of another individual
Stand over	to stand over opponent's body, or place the forepaws on the opponent and over the negative response from the opponent which is growling and trying to get out of the situation
Stand tall	an individual draws itself up to its full height; the neck is often arched and the ears pricked
T-position	one individual approaches the shoulder region of another one and often puts its head on its shoulder; formation looks like the capital "T"
Ride up	to mount another one from behind or from side

2.4 Statistical analysis

The program R 2.12.2. was used for the statistical analysis, and the graphs were made with the PAWS Statistics 18. The data could not be transformed into normal distribution, therefore statistical tests using quasibinomial distribution and poisson distribution were applied.

I analysed the distance to food as well as frequency of aggression (weak, intermediate, and strong aggression separately) with mixed effect models using a poisson distribution. I was interested whether "species" (dog vs. wolf; ecological definition of species), age (young vs. older), sex and rank (dominant vs. all other animals, that is we compared the behaviour of the highest ranking animal with the rest in each pack) had an influence on these measurements. I checked for an interaction between rank and "species". I included subject and test number as random factors in the model. Furthermore, since the observation time was not the same for all experiments the duration of each observation session was included into the model as an offset function.

Additionally, I analysed the proportion of time an animal spent with eating or showed aggression (weak, intermediate, and strong) using a mixed effect model with quasi-binomial distribution. However, if I used the proportion of time (values between 0 and 1) the distribution of the error as well was between 0 and 1. I was using the same factors as described above (except the offset function).

The dominance hierarchy for the dogs and the wolves was calculated with the David's score (Gammell et al. 2003). All the dominant and submissive behaviours which were calculated from the feeding videos were used for this (Table 5). The score is calculated with the formula:

$$DS = w + w_2 - l - l_2$$

w represented the sum of all subjects (i's) Pij values. The Pij value was the proportion of wins by individual i in his interactions with individual j. l represented the proportion of losses by i in interactions with j, $Pji = 1 - Pij$. w_2 represented the summed w values of those individuals with which i interacted, and l_2 represented the summed l values (weighted by the appropriate Pji values) of those individuals with which i interacted (Gammell et al. 2003).

With the Landau's Index the linearity of the hierarchy was checked. The result could range between 0 and 1, whereas 0 means not linear and 1 totally linear.

3 Results

3.1 Dominance hierarchy of the packs

The Landau's linearity index showed a linear dominance hierarchy for all the packs (dog pack 1: 0.6; dog pack 2: 1; wolf pack: 0.87) and with the David's score it was found that Rafiki was the highest ranking animal of the dog pack 1 (Table 6). In dog pack 2 Kilio (Table 6) and in the wolf pack Kaspar had the dominant position (Table 7).

Table 6: Dominance hierarchy of the dog packs (Table 1, 2), calculated with the David's score.

Packs	Individuals	DS	Packs	Individuals	DS
dog pack 1	Rafiki	8	dog pack 2	Kilio	1
	Maisha	2		Meru	-1
	Asali	-1			
	Bashira	-3			
	Binti	-3			
	Hakima	-3			

Table 7: Dominance hierarchy of the wolf pack (Table 3), calculated with the David's score.

Individuals	DS
Kaspar	27.6
Aragorn	24.1
Nanuk	13.4
Shima	10.5
Cherokee	-6.6
Yukon	-10.8
Apache	-11.2
Tatonga	-22.7
Geronimo	-24.4

3.2 Comparison of dogs and wolves

3.2.1 Feeding time

In the "one piece" condition we found a significant interaction between species and rank in the time the animals spent with feeding (nlme: t_{13}= 4.17, p=0.001). The highest ranking animals of each dog pack ate longer than the other pack members. In the wolf pack the dominant and the other animals ate more or less equally long

(Figure 1). In the two pieces condition there was just a difference between species (nlme: t_{13}= 8.52, p<0.001), the wolves ate longer than the dogs. In the "one piece each" condition there was an interaction of species and rank (nlme: t_{12}= 4.38, p<0.001), in the dogs the dominant animals ate longer than the others whereas in the wolves the dominant animal ate for a shorter time than the others (Figure 2). Additionally, in both species the males ate longer than the females (nlme: t_{12}= 2.20, p=0.048).

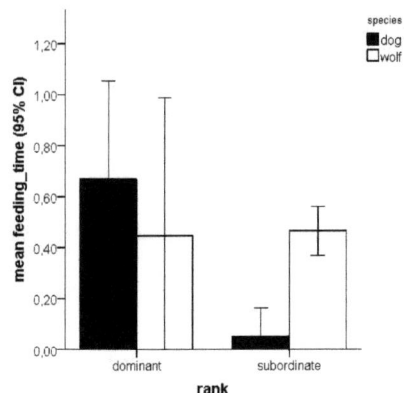

Figure 1: Feeding time (proportion of observational time) (mean and 95% confidence interval) of the dominant animals and of all other members in the wolf and dog packs in the "one piece" condition.

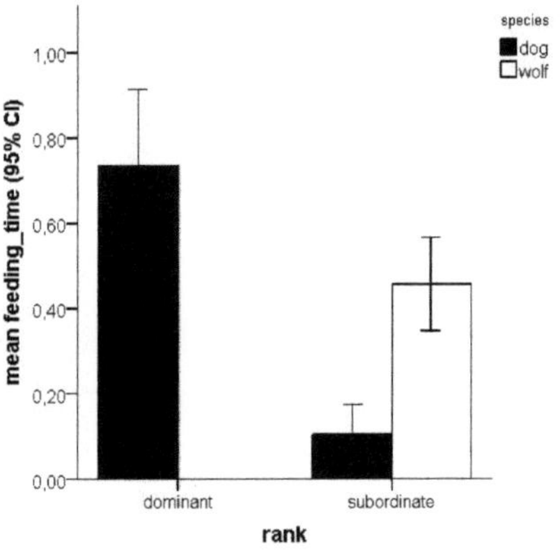

Figure 2: Feeding time (proportion of observational time) (mean and 95% confidence interval) of the dominant animals and of all other members in the wolf and dog packs in the "one piece each" condition.

Also in the "several pieces" condition we found an interaction between species and rank (nlme: t_{13}= 4.44, p<0.001). The dominant dogs ate longer, whereas in the wolves rank did not influence the length of feeding (Figure 3).

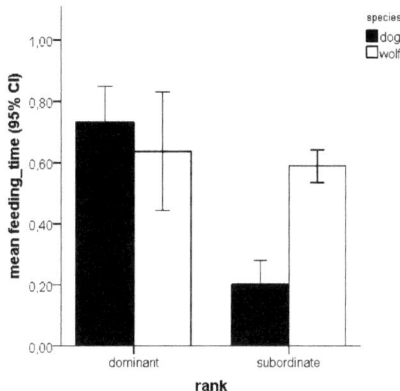

Figure 3: Feeding time (proportion of observational time) (mean and 95% confidence interval) of the dominant animals and of all other members in the wolf and dog packs in the "several pieces" condition.

3.2.2 Duration and frequency of aggressive behaviours

3.2.2.1 One piece

There were species differences in all three aggression levels in the duration (weak aggression: nlme: t_{11}= -20.47, p<0.001 [Figure 4]; intermediate: nlme: t_{12}= -16.28, p<0.001; strong: nlme: t_{13}= 26.71, p<0.001) as well as in the frequency of aggressive behaviours (weak: nlme: t_{13}= 5.58, p<0.001 [Figure 5]; intermediate: nlme: t_{12}= -21.40, p<0.001; strong: nlme: t_{11}= 48.71, p<0.001). The only variable in which the dogs appeared to be more aggressive was the duration of weak aggression. With all other 5 variables the wolves proved to be more aggressive than the dogs if only one piece of food had been provided. There was no strong aggression in dogs in the one piece condition.

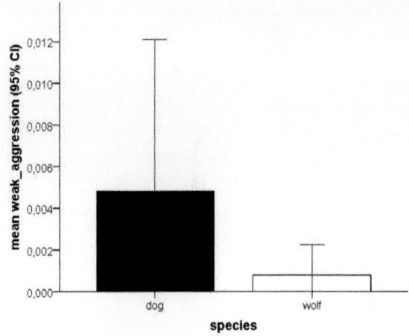

Figure 4: Duration of weak aggression (proportion of observational time) (mean and 95% confidence interval) of the wolf and dog packs in the "one piece" condition.

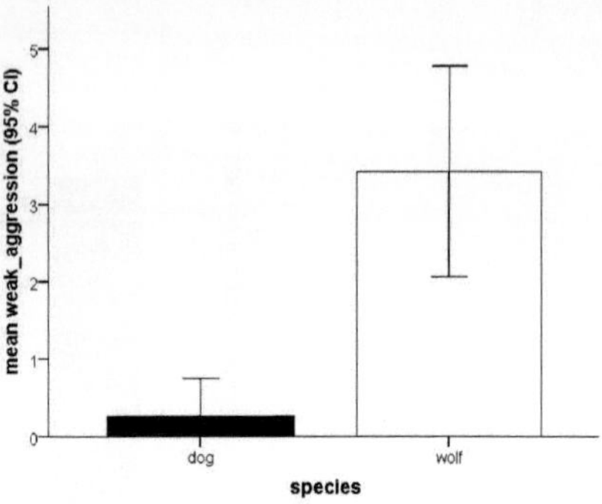

Figure 5: Frequency of weak aggression (proportion of observational time) (mean and 95% confidence interval) of the wolf and dog packs in the "one piece" condition.

In dogs the dominant animals showed significantly longer and more often weak and intermediate aggression than the subordinate ones (respectively, duration weak: nlme: t_5= -81.85, p<0.001; frequency weak: nlme: t_6= -141.78, p<0.001 [Figure 6]; duration intermediate: nlme: t_6= -57.88, p<0.001; frequency intermediate: nlme: t_6= -20.79, p<0.001). In contrast, in the wolves there was no difference between the dominant animal and the others in the amount of weak and intermediate aggression, but the subordinates showed significantly longer strong aggression than the dominant animal (nlme: t_5= 9.16, p<0.001) (Figure 7).

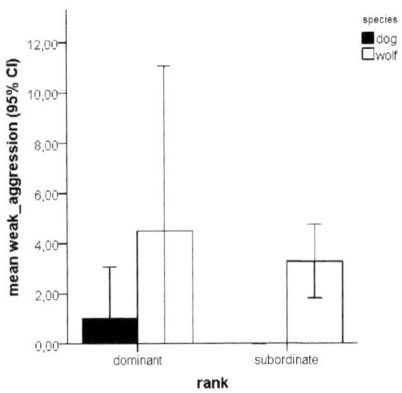
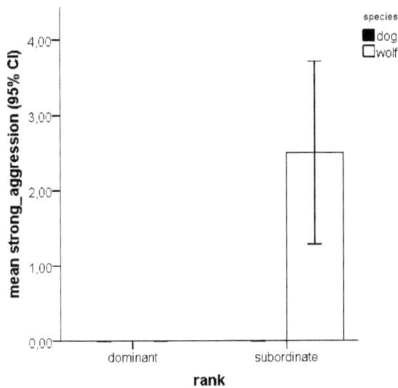

Figure 6: Frequency of weak aggression (proportion of observational time) (mean and 95% confidence interval) of the dominant animals and of all other members in the wolf and dog packs in the "one piece" condition.

Figure 7: Frequency of strong aggression (proportion of observational time) (mean and 95% confidence interval) of the dominant animals and of all other members in the wolf and dog packs in the "one piece" condition.

In the dogs no age and sex differences were found. In the wolves, the younger animals spent more time showing aggressive behaviours than the older animals (weak nlme: $t_6 = -67.59$, p<0.001; intermediate nlme: $t_6 = 2.97$, p=0.025; strong: $t_5 = 2.29$, p=0.0704) and the frequency of the intermediate and strong forms of aggression appeared also more often in the young wolves than the older ones (intermediate: nlme: $t_6 = 3.46$, p=0.0135; strong: nlme: $t_6 = 2.89$, p=0.0277). Also, in the wolves, the males spent more time with strong aggression than the females (nlme: $t_5 = 4.90$, p=0.0045) and they showed more often weak aggression than the females did (nlme: $t_6 = 2.56$, p=0.043).

3.2.2.2 Two pieces

We found species differences in all three aggression levels for the duration (weak: nlme: $t_{11} = -5.89$, p<0.001 [Figure 8]; intermediate: nlme: $t_{13} = 6.39$, p<0.001 [Figure 9]; strong: nlme: $t_{11} = 1097.34$, p<0.001) as well as for the frequency of aggression (weak: nlme: $t_{11} = -9.55$, p<0.001; intermediate: nlme: $t_{13} = 2.52$, p=0.0254; strong: nlme: $t_{13} = 57.59$, p<0.001). In each case the wolves showed longer and more frequent aggression than the dogs did. There was no strong aggression in dogs in the "two pieces" condition.

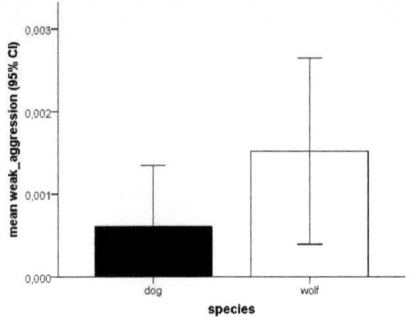 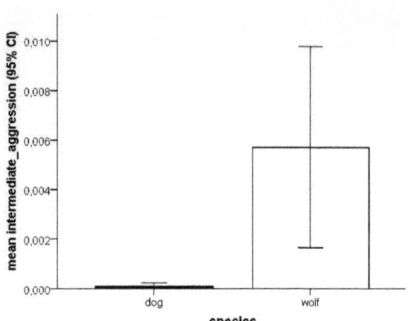

Figure 8: Duration of weak aggression (proportion of observational time) (mean and 95% confidence interval) of the wolf and dog packs in the "two pieces" condition.

Figure 9: Duration of intermediate aggression (proportion of observational time) (mean and 95% confidence interval) of the wolf and dog packs in the "two pieces" condition.

In the dogs, the dominant animals showed weak aggression longer (nlme: t_4= -289.13, p<0.001) (Figure 10) and more often (nlme: t_4= -17.68, p<0.001) than the other pack members did (Figure 11). In the wolves the dominant animal did not differ compared to the others.

The male wolves, however, showed significantly longer strong aggression than the females (nlme: t_6= 3.26, p=0.0173) whereas in the dogs no sex difference was found. On the contrary, in the wolves the two age groups did not differ, while in the dogs, the older animals showed more frequent intermediate aggressive behaviours than the young ones (nlme: t_4= 8.31, p=0.0011).

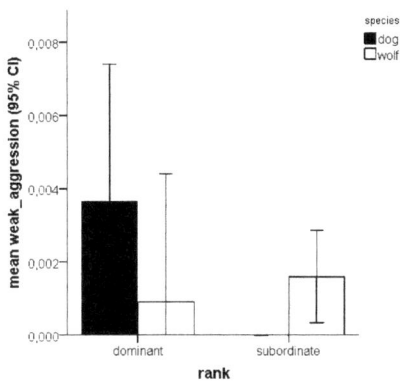 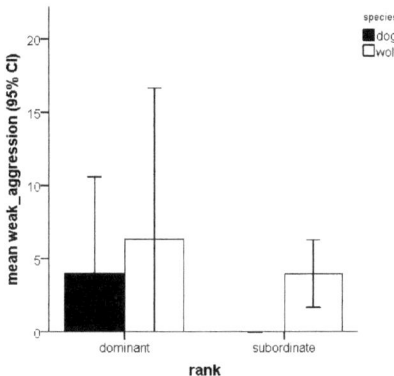

Figure 10: Duration of weak aggression (proportion of observational time) (mean and 95% confidence interval) of the dominant animals and of all other members in the wolf and dog packs in the "two pieces" condition.

Figure 11: Frequency of weak aggression (proportion of observational time) (mean and 95% confidence interval) of the dominant animals and of all other members in the wolf and dog packs in the "two pieces" condition.

3.2.2.3 One piece each

Again we found a species difference in the duration of weak (nlme: t_{13}= -3.38, p=0.005) (Figure 12) and strong aggression (nlme: t_{13}= -2.40, p=0.0323) and in the frequency of intermediate (nlme: t_{14}= 2.57, p=0.0222) (Figure 13) and strong aggression (nlme: t_{12}= -6.50, p<0.001). The dogs spent more time with weak aggression than the wolves did whereas the wolves showed longer and more frequent strong aggression and spent more time also with intermediate aggression.

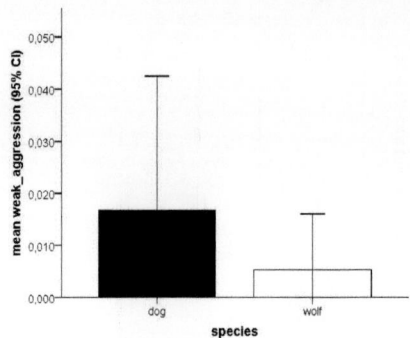

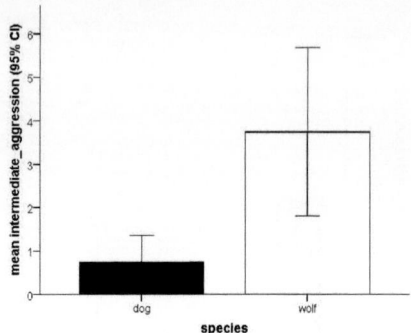

Figure 12: Duration of weak aggression (proportion of observational time) (mean and 95% confidence interval) of the wolf and dog packs in the "one piece each" condition.

Figure 13: Frequency of intermediate aggression (proportion of observational time) (mean and 95% confidence interval) of the wolf and dog packs in the "one piece each" condition.

A rank difference was detected only in the dogs. The dominant animals showed longer and more frequent weak aggression than the subordinate ones (duration nlme: t_3= -21.89, p<0.001 [Figure 14]; frequency nlme: t_6= -3.49, p=0.013 [Figure 15]) and a tendency into the same direction in strong aggression as well (duration nlme: t_6= -2.24, p=0.0661; frequency nlme: t_6= -2.29, p=0.0617).

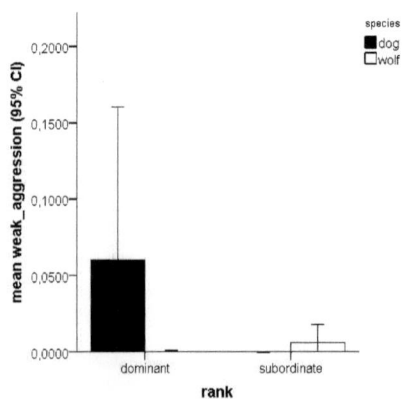

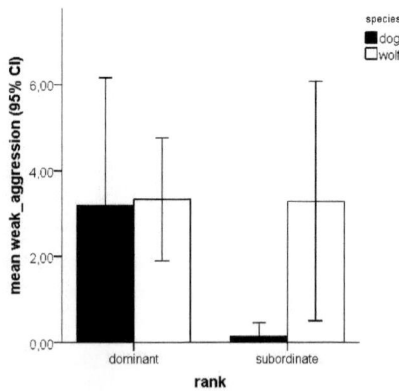

Figure 14: Duration of weak aggression (proportion of observational time) (mean and 95% confidence interval) of the dominant animals and of all other members in the wolf and dog packs in the "one piece each" condition.

Figure 15: Frequency of weak aggression (proportion of observational time) (mean and 95% confidence interval) of the dominant animals and of all other members in the wolf and dog packs in the "one piece each" condition.

Regarding sex differences, we only found that the male dogs spent more time with weak aggression than the females did (nlme: t_3= 6.23, p=0.0083), and that the female wolves showed intermediate aggression longer than the males (nlme: t_5= -5.63, p=0.0025).

In the dogs, the older animals spent more time with weak aggression (nlme: t_3= 28.58, p<0.001). In the wolves, the older animals showed more often weak and intermediate aggression than the younger ones (nlme: t_7= -5.84, p<0.001 and t_7= -2.57, p=0.0371, respectively) but the young wolves were intermediately aggressive for a longer time (nlme: t_5= 5.67, p=0.0024).

3.2.2.4 Several pieces

The dogs spent more time with weak (Figure 16) as well as strong aggression than the wolves did (nlme: t_{12}= -13.17, p<0.001 and t_{13}= -200.58, p<0.001, respectively) but the wolves showed more often weak aggression (nlme: t_{14}= -5.35, p<0.001) (Figure 17) and more often (nlme: t_{12}= -29.20, p<0.001) and longer (nlme: t_{13}= -13.49, p<0.001) intermediate aggression than the dogs did.

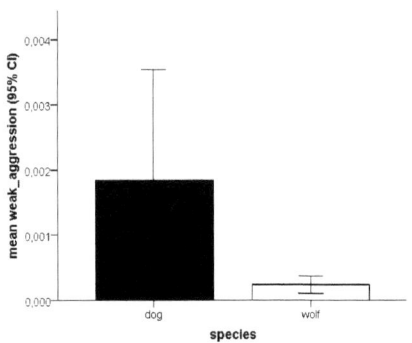

Figure 16: Duration of weak aggression (proportion of observational time) (mean and 95% confidence interval) of the wolf and dog packs in the "several pieces" condition.

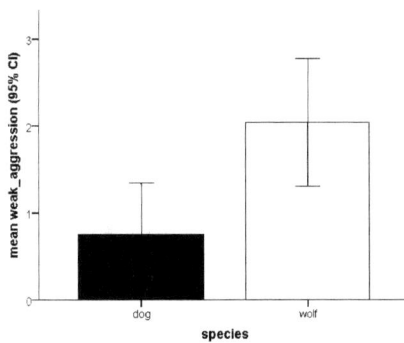

Figure 17: Frequency of weak aggression (proportion of observational time) (mean and 95% confidence interval) of the wolf and dog packs in the "several pieces" condition.

Only in the dogs a rank effect was found. The dominant dogs showed longer aggression at the weak (nlme: t_6= -31.71, p<0.001) and intermediate (nlme: t_6= -

34.83, p<0.001) levels and more often aggression in all three aggression levels (weak nlme: t_6= -6.10, p=0.0017; intermediate nlme: t_5= -26.47, p<0.001; strong nlme: t_3= -8.08, p=0.004) than the other members of the packs did.

A significant sex difference was detected only in the dogs and only regarding strong aggression. The males showed longer strong aggression (nlme: t_5= -574.40, p<0.001) and also more often (nlme: t_3= -52.94, p<0.001) (Figure 18). In dogs, the older animal showed more often weak aggression (nlme: t_5= -16.92, p<0.001) (Figure 19) and also longer (nlme: t_5= -130.36, p<0.001), and more often (nlme: t_3= 4.19, p=0.0247) strong aggression.

In wolves, the older members of the pack showed more often weak aggression (nlme: t_7= -3.27, p=0.0136) (Figure 19) and there was also a strong tendency for more intermediate aggression found in wolves (nlme: t_7= -2.31, p=0.0543).

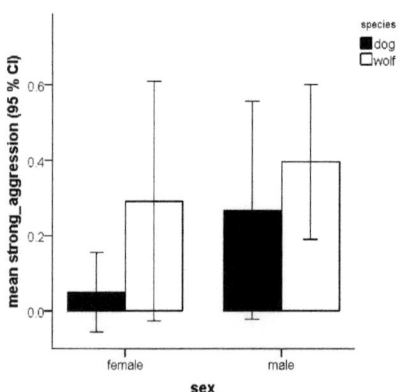 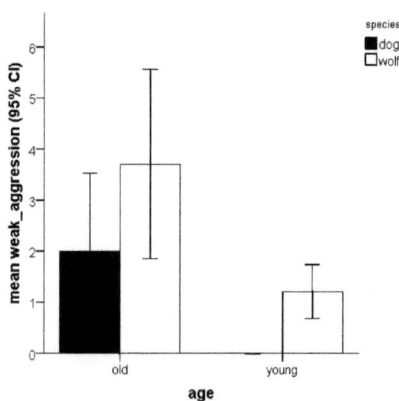

Figure 18: Frequency of strong aggression (proportion of observational time) (mean and 95% confidence interval) of the female and male animals in the wolf and dog packs in the "several pieces" condition.

Figure 19: Frequency of weak aggression (proportion of observational time) (mean and 95% confidence interval) of the older and younger animals in the wolf and dog packs in the "several pieces" condition.

Table 8: Summary of the statistically confirmed differences in the duration and frequency of aggression in dogs and wolves.

		one piece		two pieces		one piece each		several pieces	
		duration	frequency	duration	frequency	duration	frequency	duration	frequency
species	weak aggression	D > W	D < W	D < W	D < W	D > W	D = W	D > W	D < W
	intermediate aggression	D < W	D < W	D < W	D < W	D = W	D < W	D < W	D < W
	strong aggression	- < W	- < W	- < W	- < W	D < W	D < W	D > W	D = W
age WOLF	weak aggression	y > 0	y = 0	y = 0	y = 0	y = 0	y < 0	y = 0	y < 0
	intermediate aggression	y > 0	y > 0	y = 0	y = 0	y > 0	y < 0	y = 0	y </= 0
	strong aggression	y >/= 0	y > 0	y = 0	y = 0	y = 0	y = 0	y = 0	y = 0
age DOG	weak aggression	y = 0	y = 0	y = 0	y = 0	y < 0	y = 0	y = 0	y < 0
	intermediate aggression	y = 0	y = 0	y = 0	y < 0	y = 0	y = 0	y = 0	y = 0
	strong aggression	-	-	-	-	y = 0	y = 0	y < 0	y < 0
sex WOLF	weak aggression	f = m	f < m	f = m	f = m	f = m	f = m	f = m	f = m
	intermediate aggression	f = m	f = m	f = m	f = m	f > m	f = m	f = m	f = m
	strong aggression	f < m	f = m	f < m	f = m	f = m	f = m	f = m	f = m
sex DOG	weak aggression	f = m	f = m	f = m	f = m	f < m	f = m	f < m	f = m
	intermediate aggression	f = m	f = m	f = m	f = m	f = m	f = m	f = m	f = m
	strong aggression	-	-	-	-	f = m	f = m	f < m	f < m
dom WOLF	weak aggression	dom = others	dom = others	dom = others	dom = others	dom = others	dom = others	dom = others	dom = others
	intermediate aggression	dom = others	dom = others	dom = others	dom = others	dom = others	dom = others	dom = others	dom = others
	strong aggression	dom < others	dom = others	dom = others	dom = others	dom = others	dom = others	dom = others	dom = others
dom DOG	weak aggression	dom > others	dom > others	dom > others	dom > others	dom > others	dom > others	dom > others	dom > others
	intermediate aggression	dom > others	dom > others	dom = others	dom = others	dom = others	dom = others	dom > others	dom > others
	strong aggression	-	-	-	-	dom >/= others	dom >/= others	dom > others	dom > others

(D = Dogs, W = Wolves, y = younger animals (basically the second generation), o = older animals (basically the first generation), f = females, m = males, dom = dominant animal(s), others = the other animals in the pack, basically the subordinates; > the ones on the left showed longer aggression in the duration columns or showed more aggression in the frequency columns; < the same as before just now the longer and more aggression is on the right side; = means that there was no significant difference in the aggression; - means that there was no aggression found)

The results seem to indicate that in most cases the wolves showed aggressive behaviours in a stronger form, more often and for a longer time than the dogs did, though this gradually changed as more and more pieces of food were available in a distributed form. Also there was a clear difference between dogs and wolves from the point of view that in the dogs the highest ranking animals behaved more aggressively that the other pack members whereas in the wolves no such difference appeared. This seems to be in line with the species differences we found in the time the dominant and other animals spent with eating. In the wolves being a dominant or not did not influence the feeding time in most conditions whereas in the dog packs the dominant animals were feeding for a longer time that the others. Based on these results, it seems possible that in dogs less aggression occurred because the dominant animals often successfully monopolized the food and in these cases the other animals did not even approach the food. That is, in dogs there might have been less reason for agonistic encounters than in wolves. To check for this opportunity we measured how much time the animals spent in which distance to the food.

3.2.3 Distance to food

3.2.3.1 One piece

An interaction between species and rank was found for all three distance-to-food categories (next nlme: t_{12}= -2.457, p=0.0302; close nlme: t_{13}= 2.758, p=0.0163; distant nlme: t_{12}= -4.461, p<0.001) (Figure 20). In the dogs the dominant animals spent more time directly next and close to the food than the other pack members did and the subordinate dogs spent more time distant to food. In the wolves, there was just a difference found in distant to food, the dominant wolves spent more time there compared to the subordinate ones.

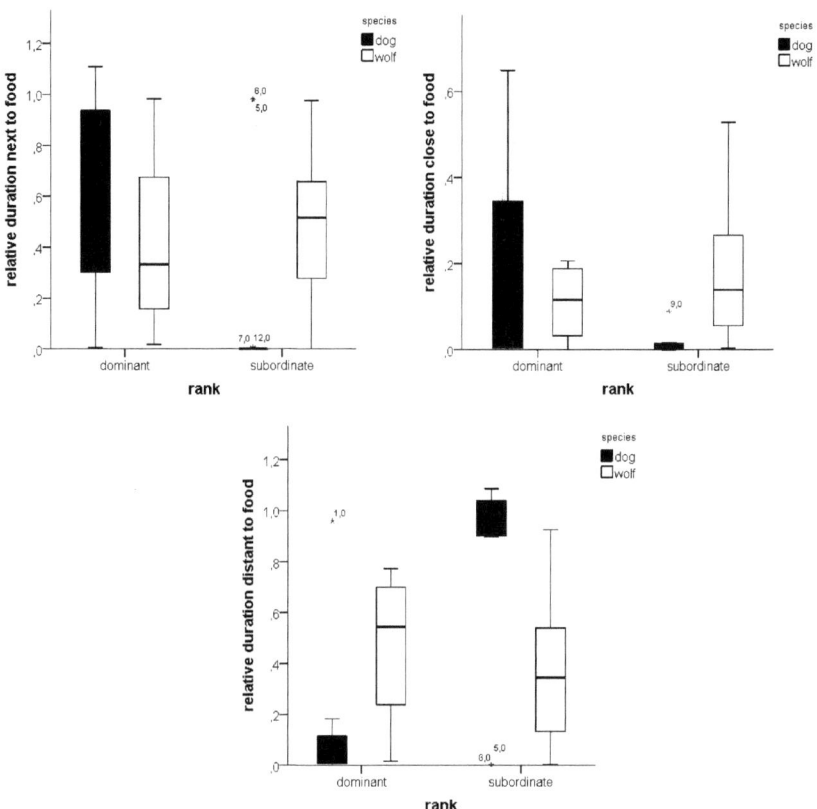

Figure 20: Time spent next, close and distant to food (proportion of observational time) (mean and 95% confidence interval) of the dominant animals and of all other members in the wolf and dog packs in the "one piece" condition.

3.2.3.2 Two pieces

In the next (nlme: t_{13}= 3.32, p=0.0055) and distant (nlme: t_{12}= -4.96, p<0.001) categories there was an interaction between species and rank found. In the dogs, the dominant animals spent more time next to food and the subordinate ones distant to food. In wolves, there was no rank difference found in next and distant to food. In the close to food category there was just a species difference, the wolves spent more time there (nlme: t_{14}= 6.94, p<0.001) (Figure 21).

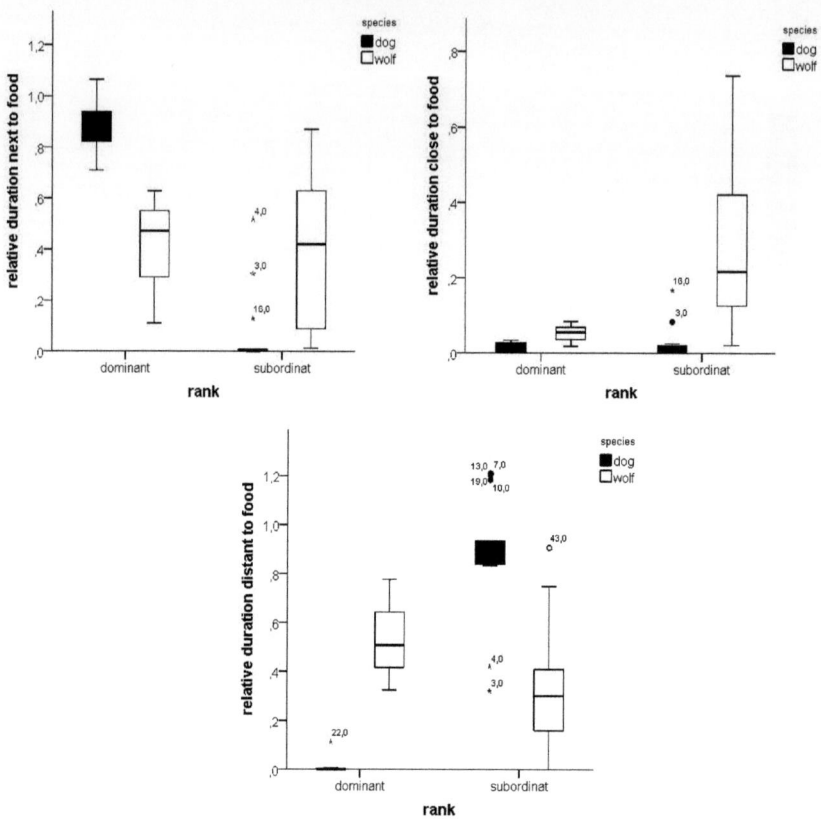

Figure 21: Time spent next, close and distant to food (proportion of observational time) (mean and 95% confidence interval) of the dominant animals and of all other members in the wolf and dog packs in the "two pieces" condition.

3.2.3.3 One piece each

An interaction between species and rank were found in the amount of time spent next (nlme: $t_{11}= 6.876$, $p<0.001$) and distant (nlme: $t_{13}= -5.915$, $p<0.001$) to food. The dominant dogs spent more time next to the food and the subordinate dogs distant to food. In the wolves the subordinates spent more time next as well as distant to food than the dominant animals. In close to food there was just a species difference found (nlme: $t_{14}= 4.248$, $p<0.001$), the wolves spent more time there than the dogs (Figure 22).

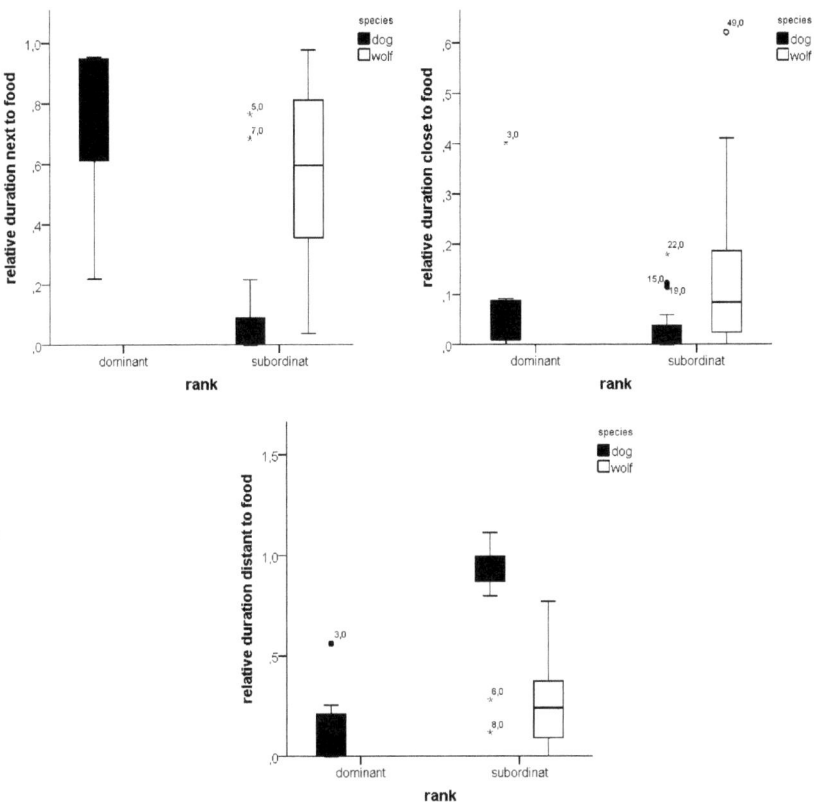

Figure 22: Time spent next, close and distant to food (proportion of observational time) (mean and 95% confidence interval) of the dominant animals and of all other members in the wolf and dog packs in the "one piece each" condition.

3.2.3.4 Several pieces

An interaction between species and rank was found in the variables time spent next (t_{13}= 3.617, p=0.0031) and distant (t_{13}= -3.430, p=0.0045) to food. In close to food there was just a rank difference found (t_{15}= 7.321, p<0.001). The dominant dogs spent more time next to food and the subordinate in close and distant. In the wolves no rank difference was found in next to food but the subordinate animals spent more time close to food and the dominant animal distant to food (Figure 23).

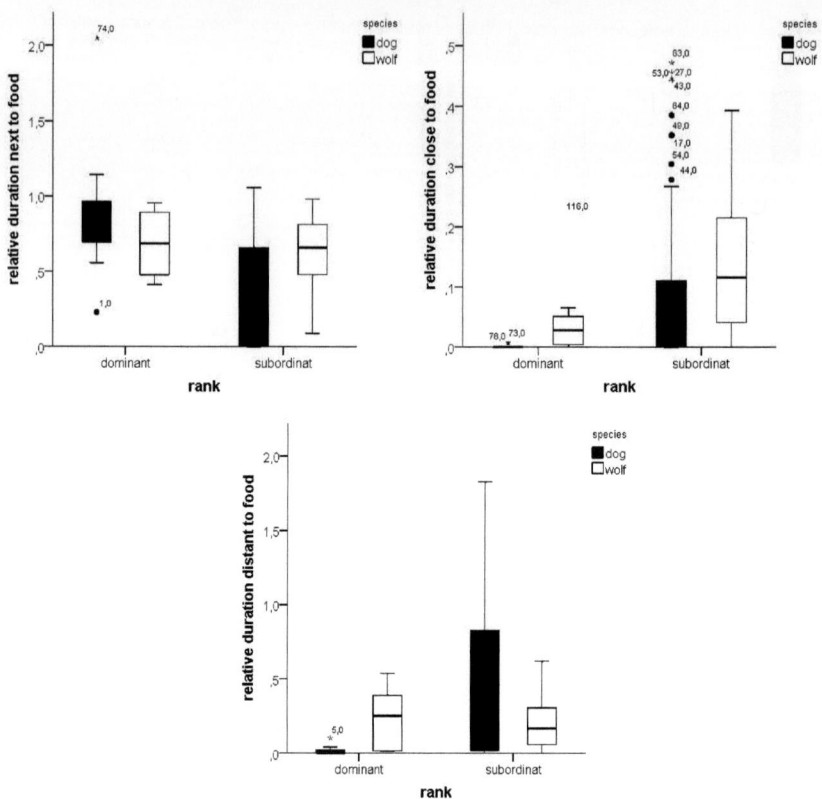

Figure 23: Time spent next, close and distant to food (proportion of observational time) (mean and 95% confidence interval) of the dominant animals and of all other members in the wolf and dog packs in the "several pieces" condition.

3.3 Summary of the results

In the previous section we found that in the two dog packs the highest ranking animals were more aggressive than the other members of the pack and they successfully monopolized the food. This was shown by the fact that the dominant dogs ate for a longer time than the subordinate ones. In the wolves no such effects were found since dominant and subordinate wolves spent a comparable amount of time with eating and were similarly aggressive. By further analysing how much time the animals spent next, close and distant to food, we demonstrated that in the dogs the subordinate animals did not even try to approach the food. Mostly, the dogs spent less time standing close to the food without actually having it, in comparison to the wolves. Instead, the dominant dogs spent more time next to the food (eating or guarding it) than the subordinates did who, however, spent more time distant to food than the dominant dogs did. This was true in all 4 conditions. Only in the several pieces condition the subordinate dogs spent more time close to the food than the dominant dogs did – only in this context their behaviour was somewhat similar to that of the wolves.

4 Discussion

4.1 Comparison of dogs and wolves

The aim of this study was to find out if there is a difference in different feeding situations between dogs and wolves, which were raised and housed similarly in captive packs. Unfortunately the sample size was very small, only one wolf pack and two dog packs, and therefore any interpretations must be treated with caution.

We found out that

1) in total the wolves showed more aggression than dogs. This seems to support the emotional reactivity hypothesis. When looking at the details, however, this conclusion may be premature.

2) Indeed, according to the data wolves spent more time with intermediate and strong aggression than the dogs did in most condition and also the frequency of weak aggression was higher. The dogs, however, spent more time with weak aggression in most test conditions, showing that such agonistic interactions went on longer in the dogs than in the wolves.

3) Further on, when looking at not only the amount of aggressive behaviours but also who had access to the food and whether the animals were at all close to the resource, we found important differences between dogs and wolves. In most feeding conditions the dominant and all other wolves spent a similar amount of time close to the food. Only one condition was an exception, where actually the subordinate animals were longer in proximity to the food than the dominant wolf. On the contrary, in the dogs the alpha male was significant longer near the food than the rest of the pack who were typically not within range.

4) This seems to indicate a steeper dominance hierarchy in the dogs than in the wolves. This is further confirmed by finding that most of the time, the dominant dogs showed significant more and longer aggression in all three aggression levels than the other dogs did. Whereas in the wolves, all animals showed the similar amount and

time of aggression, just one time the subordinate wolves showed even longer aggression.

In the two conditions with the fewest pieces the wolves showed longer and more aggression in all levels except of the duration of weak aggression with just one piece, there the dogs showed longer aggression. In the conditions with more pieces there was more diversity. In both conditions with distributed food, the dogs showed weak aggression longer than the wolves did but the wolves showed such behaviours more often. In the intermediate aggression level the wolves showed longer and more aggression than the dogs as well as in the strong aggression for the condition where everybody had a piece. In the monopolizable condition the dogs showed longer aggression and the frequency was the same in both, dogs and wolves.

Higher aggression in monopolizable food distribution than in highly shareable one was observed also in a lot other species, like chimpanzees (Hare et al. 2007), gorillas *(Gorilla gorilla)* (Scott & Lockard 2006), white-faced capuchin monkeys (*Cebus capucinus*) (Vogel, Munch & Janson 2007) and Japanese medaka *(Oryzias latipes)* (Robb & Grant 1998).

So far, the results showed that the wolves showed more aggression than the dogs but on the feeding time we can see that in wolves all animals had access to the food and in the dogs the dominant animals ate significant longer than the subordinates which implicates that the dominants had more access to the food. This was visible also in the time dominant and subordinate animals spent close to food. In dogs the dominant animals were more often next to the food which included also feeding and the subordinate animals were most of the time distant to the food. In the wolves all animals were either the same amount of time at all distances or the subordinate were longer next and close to the food and the dominant ones distant.

In a lot of hierarchical ordered animal species it is common that the dominant animals are allowed to eat first like it was in the dogs. For example in long-tailed macaques *(Macaca fascicularis)* (van Schaik & van Noordwijk 1988, Dubuc & Chapais 2007), chimpanzees (Hare et al. 2007), ringtailed lemurs *(Lemur catta)* (Nunn & Deaner

2004) and domesticated chickens *(Gallus gallus domesticus)* (Schjelderup-Ebbe 1935).

According to the results of the feeding time and distance to food it seemed that the wolves are more tolerant than the dogs because in the dogs just the dominant animals ate and spent more time next and close to the food. Thus, it is likely that the lower aggression of the dogs is due to the lack of confrontation. Another reason why the wolves appeared in total more aggressive was that in the wolves more animals showed aggression than in the dogs where almost only the dominant initiated aggressive encounters and the others fled without fighting. In the wolves there was no difference between the ranks all animals showed about the same duration and frequency of aggression. In contrast the dominant dogs showed nearly always longer and more often aggression than the other members of the packs. Accordingly, one can easily explain the lower levels of aggression in dogs compared to wolves by the fact that the subordinate dogs seemed to try to avoid conflicts with the highest ranking member of the pack. When the subordinate dogs, which are the majority of the pack, are not near the food there is no need for the dominant to show aggression. In contrast the majority of the wolf pack was most of the time in the same distance to the food and therefore there was more reason to go into aggressive interactions. That animals show more aggression when they are close to each other was already shown in 1956 by Marler. He found out that Chaffinches (*Fringilla coelebs*) fight more often with each other when they ate close together than when their feeding places were farer apart from each other. If tolerance is defined as allowing the others come close to the food and more animals eating together, wolves appear more tolerant than dogs. In sum, our results seem to confirm the "Relaxed selection" hypothesis (e.g. Zimen 1992).

So far only indirect attempts have been made to compare wolves and dogs in their aggression and tolerance for example behaviour observations. Mech (1994), Mech et al. (1998) and Murray et al. (2010) reported about lethal intergroup conflicts in wolves, whereas Boitani et al. (1995) and Pal et al. (1999) argued that such things were not happening in feral dogs, they would tend to only bark at intruders at a distance until they leave. Hare and colleagues (2012) claimed also that the within-group aggression is also reduced in dogs compared to wolves. Dominant female wolves often try to suppress the subordinate females that they cannot breed or even attack or kill the pups of them (McLeod 1990; Sands & Creel 2004). In dogs no such behaviour is reported. Furthermore dogs tolerate the inspection of their anogenital region by another dog, but wolves even don't like it when pack members are trying this (Bradshaw & Nottingham 1995). Some general rules of aggression in wolf packs have been described whereas a lot less is known about dog packs.

The wolves might show longer or more aggression but their aggression is mostly ritualized, that is, they compete over resources with actions that have no severe consequences for either party. During this study and according to many other studies (Mech 1970, Zimen 1990, Feddersen-Petersen 1991) strong aggression in wolves was observed but never aggression that escalated into attacks or fights within the pack. In contrast, this was observed indeed in the dogs at the WSC and also in other studies (Feddersen-Petersen 1991, Zimen 1992).

Another phenomenon that regulates competition over food is the so-called "respect of possession" that has been described in wolves (Mech 1970), macaques (Kummer & Cords 1991) and baboons (Sigg & Falett 1985) and could also be observed in this study. A wolf respected the ownership of food in the mouth of another animal but if it left the piece alone just for a few seconds to drive another pack member away than the third took its advantage and stole the food (see also Mech 1970, p. 71). Even the "provocative" presentation of food in front of a dominant animal of low-ranking wolves (Mech 1999) could be observed in this study.

In dogs no information on the "respect of possession" was found. In this study it seemed that such a rule did not exist. If a subordinate managed to take a piece of food, the alpha immediately tried to get it back even with force when he noticed it. Only once in dog pack 2, which consists of only two animals, the subordinate managed to get to one piece at first and the dominant one did not take it from him the whole time. He however stayed close to the subordinate dog eating, and whined and ran around him. Nevertheless a lot more work is needed to study such interactions of dogs in various contexts in order to learn about the rules of their social life. Actually, also the context of the situation is very important, it would be important to have a look at different groups living in different conditions to find out more about this phenomenon.

These differences between dogs and wolves could be based on epigenetics. This assumption was also made by Thierry (1997) that individuals affect each other through social contacts. The weaker opponent in a conflict will submit or flee when the risk of being injured is high (Parker 1974). However, when the dominance gradient is low the intensity of aggression tends to be higher and the asymmetry of contests is strong (Preuschoft & van Schaik 2000).

Epigenetic shows that the environment can have a big influence on the personality of an individual. It can adjust genetic predominated behaviour pattern and form new ones. Maybe there was more aggression in wolves than in dogs because the subordinate dogs had learned to avoid the food when the alpha is present because then they will receive aggression and therefore they stayed away (also explained by Miklósi 2007, p. 21).

In the wolves in this study there was no such behaviour observed. On the contrary, the younger animals often tried to sneak closer or come closer with submissive behaviour to another pack member with food. Apparently, they tried to calm the other wolf down in order to co-feed with it (personal observation). Especially the youngest of the observed animals often succeeded with this tactic. The puppies are those who are lowest in rank (Mech 1970, Pal et al. 1998). Their excessive and fierce behaviour toward older group members should serve as pacification of aggression and trigger care behaviour (Zimen 1990). "Canine behavior toward puppies appears to be governed by a social code that forbids injurious bites or life-threatening attacks" (Lindsay 2005). Additionally there are also references of the behaviour of the wolves in this study. In a lot of species the adults, and therefore also the dominant ones, allow the young members of their group to eat first (e.g. wild dogs *(Lycaon pictus)*: Malcom and Marten 1982; spotted hyenas *(Crocuta crocuta)*: Frank 1986; several primates: Hand 1986; domestic cat *(Felis catus)*: Bonanni et al. 2007). Maybe a connection of this and in general more tolerance in feeding situations can explain that all wolves ate equally long or in the condition where each wolf got a piece indeed the subordinate ate longer than the dominant one.

Additionally in a lot of species it is normal that the mother tolerate that her offspring take food from her because they can gain information about what they can eat and the value of the food (Jaeggi et al. 2008, Ueno & Matsuzawa 2004) and sometimes it reduces the time to weaning (Brown et al. 2004). However in all four ape species as well as in a few monkey species, especially in callitrichids, also other adult members of the group share food with the young ones (reviewed by Brown et al. 2004). In common marmosets *(Callithrix jacchus)* the parents even differentiate between younger and older offspring. They tend to be more tolerant toward younger offspring (until 15 weeks old) than older ones (more than 29 month) (Saito et al. 2008).

It seems that this behaviour pattern does not work anymore in the dogs. One reason might be that nowadays the humans take over the responsibility of the dogs live and therefore also of the puppies. The human is determining what the dogs are eating and how they should behave. Besides in feral dogs the time when the young dogs are able to breed and disperse is much earlier than in the wolves, namely with 6 to 12 month and not with 2 years like the wolves (Morey 1994). It seems that the development of the dog is accelerated and therefore the time period for intensive care-taking of the dog puppies is much shorter.

4.2 Limitations of the study

Since I could just test one wolf and two dog packs the personality of the dominants or individual features of the packs can have a strong influence on the results. I cannot say for sure that the behaviour of these animals is typical for the whole species. A bigger sample size and more repetitions would be necessary to see if the results of this study could be transferred to dogs and wolves as a whole or if it is just true for this special sample.

The animals in this study were kept in special circumstances. The humans had a big influence on them because of the handraising and daily contact. Furthermore all the animals participated in other behavioural and cognition tests since they were four weeks old, what might have an influence also on the general behaviour of them.

Another possible explanation for the behaviour of this dog pack could be that they didn't fight with the alpha of the pack because they anyway get food (mostly dry food) from the humans, because they lived in a research centre and therefore participated also in other studies at this time. Because at the Wolf Science Centre the wolves were not fed every day with carcasses and the dogs were fed every day with dry food. The reason for that is just because it is more natural for wolves if they were not fed every day but during the domestication the digestive tract of the dog changed and its better for them to feed them every day.

And due to time limits there was maybe to less time between the single feeding sessions and they were not motivated enough. Thus, it would be better to prolong the time between the feedings for the dogs and do not feed them much in between to keep the motivation high to get food.

Another constraint of this study could be that the number of animals in the packs is different. With 9 animals in the wolf pack and 6 respectively 2 in the dogs the number is quite different.

Additionally a reduction of the test conditions could be worth to do. I think one big piece compared with pieces would be as meaningful as the version in this study but it would be easier to compare and therefore clearer and more focused.

4.3 Conclusion

However, in the end I think that both hypotheses are true: they are likely to describe evolutionary processes that influenced the behaviour of dogs, but in different areas of their social life. I do not think that either the wolves or the dogs are more aggressive in a general way. The wolves are more tolerant within the pack, they need each other to survive and therefore it would be self-defeating to harm the pack members. For this reason wolves have ritualized their behaviour to form and stabilize their hierarchy with the minimum loss of energy (Mech 1970, Miklósi 2007). In contrast wolves are extremely aggressive to strange wolves and many reports exist that they kill unfamiliar wolves when they enter their territory (e.g. Mech 1970, Peterson & Ciucci 2003).

Dogs were domesticated to live and communicate with humans. A different species determined the life of dogs, when and what they eat, when they should reproduce and so on. The dog was made to be dependent on the human and not on its conspecifics. Therefore they care not so much if another dog gets something to eat. But the human selected him to be non aggressive to other also non-familiar humans, not only to his "pack" but also to be tolerant to strange dogs (Zimen 1992).

I think both strategies are perfect for their habitats and in their way both hypotheses are true.

5 References

Archer, J. 1988. The behavioural biology of aggression. University press, Cambridge, UK.

Belyaev, D. 1979. Destabilizing selection as a factor in domestication. Journal of Heredity 70: 301–308.

Berman, M. & Dunbar, I. 1983. The social behaviour of free-ranging suburban dogs. Applied Animal Ethology 10: 5-17.

Boitani, L. & Ciucci, P. 1995. Comparative social ecology of feral dogs and wolves. Ethology, Ecology & Evolution 7: 49-72.

Boitani, L., Francisci, F., Ciucci, P. and Andreoli, G. 1995. Population biology and ecology of feral dogs in central Italy. In: The domestic Dog: its evolution, behaviour and interactions with people, ed. J. Serpell. Cambridge University Press, UK. pp. 218-244.

Bonanni, R., Valsecchi, P. & Natoli, E. 2010. Pattern of individual participation and cheating in conflicts between groups of free-ranging dogs. Animal Behaviour 79: 957-968.

Bonanni, R, Cafazzo, S., Fantini, C., Pontier, D. & Natoli, E. 2007. Feeding-order in an urban feral domestic cat colony: relationship to dominance rank, sex and age. Animal Behaviour 74: 1369-1379.

Bradshaw, J. & Nottingham, H. 1995. Social and communication behavior of companion dogs. In: The Domestic Dog: Its Evolution, Behavior, and Interactions with People, ed. J. Serpell. Cambridge University Press, UK. pp. 115-130.

Bräuer, J. Kaminski, J., Riedel, J., Call, J. & Tomasello, M. 2006. Making Inferences About the Location of Hidden Food: Social Dog, Causal Ape. Journal of Comparative Psychology 120: 38-47.

Brown, J. L. 1964. The evolution of diversity in avian territorial systems. The Wilson Bulletin 76: 160-169.

Brown, G. R., Almond, R. E. A. & van Bergen, Y. 2004. Begging, Stealing, and Offering: Food Transfer in Nonhuman Primates. Advances in The Study of Behaviour 34: 265-295.

Burkart, J. M. Fehr, E., Efferson, C. and van Schaik, E. P. 2007. Other-regarding preferences in a non-human primate: Common marmosets provision food altruistically. PNAS 104: 19762-19766.

Cafazzo, S., Valsecchi, P., Bonanni, R. & Natoli, E. 2010. Dominance in relation to age, sex, and competitive contexts in a group of free-ranging domestic dogs. Behavioral Ecology 21:443-455

Clutton-Brock, J. 1995. Origins of the dog: domestication and early history. In: The domestic Dog: its evolution, behaviour and interactions with people, ed. J. Serpell. Cambridge University Press, UK. pp. 7-20.

Coppinger, R. & Coppinger, L. 2001. Dogs: A Startling New Understanding of Canine Origin, Behavior & Evolution. The University of Chicago Press: Chicago, USA.

Cronin, K. A., Schroeder, K. K. E., Rothwell, E. S., Silk, J. B. and Snowdon, C. T. 2009. Cooperatively Breeding Cottontop Tamarins (*Saguinus Oedipus*) Do Not Donate Rewards to Their Long-Term Mates. Journal of Comparative Psychology 123: 231-241.

Daniels, T. J. & Bekoff, M. 1989. Population and Social Biology of Free-Ranging Dogs, *Canis familiaris*. Journal of Mammalogy 70: 754-762.

Davis, S. J. & Valla, F. R. 1978. Evidence for domestication of the dog 12,000 years ago in the Natufian of Israel. Nature 276: 608-610.

de Waal, F. B. M. 1997. Food Transfers Through Mesh in Brown Capuchins. Journal of Comparative Psychology 111: 370-378.

Dubuc, C. & Chapais, B. 2007. Feeding Competition in *Macaca fascicularis*: An Assessment of the Early Arrival Tactic. International Jouranl of Primatolgy 28: 357-367.

Feddersen-Petersen, D. U. 1991. The ontogeny of social play and agonistic behaviour in selected canid species. Bonner zoologische Beiträge 42: 97-114.

Feddersen-Petersen, D. U. 2004. Hundepsychologie. Sozialverhalten und Wesen. Emotionen und Individualität. Franckh-Kosmos Verlag-GmbH & Co. KG, Stuttgart, Germany. pp. 294-307.

Frank, H. 1980. Evolution of Canine Information Processing under Conditions of Natural and Artificial Selection. Zeitschrift für Tierpsychologie 53: 389-399.

Frank, L. G. 1986. Social-Organization of the Spotted Hyena *(Crocuta crocuta)*. 2. Dominance and Reproduction. Animal Behaviour 34: 1510-1527.

Frank, H. & Frank, M. G. 1982. On the effects of domestication on canine social development and behavior. Applied Animal Ethology 8: 507-525.

Furuichi, T. 2011. Female Contributions to the Peaceful Nature of Bonobo Society. Evolutionary Anthropology 20: 131-142.

Gácsi, M., Győri, B., Virányi, Z., Kubinyi, E., Range, F., Belényi, B. & Miklósi, Á. 2009. Explaining Dog Wolf Differences in Utilizing Human Pointing Gestures: Selection for Synergistic shifts in the Development of Some Social Skills. PLoS ONE 4: e6584

Gammell, M. P., de Vries, H., Jennings, D. J., Carlin, C. M. & Hayden, T. J. 2003. David's score: a more appropriate dominance ranking method than Clutton-Brock et al.'s index. Animal Behaviour 66: 601-605.

Germonpré, M., Sablin, M. V., Stevens, R. E., Hedges, R. E. M., Hofreiter, M., Stiller, M. & Després, V. R. 2009. Fossil dogs and wolves from Palaeolithic sites in Belgium, the Ukraine and Russia: osteometry, acient DNA and stable isotopes. Journal of Archaeological Science 36: 473-490.

Goodmann, P. A. & Klinghammer, E. 1990. Wolf ethogram. North American Wildlife Park Foundation (Wolf Park), Battle Ground, Indiana

Goodwin, D. Bradshaw, J. W. & Wickens, S. M. 1997. Paedomorphosis affects agonistic visual signals of domestic dogs. Animal Behaviour 53: 297-304.

Gray, M. M. & Wayne, R. K. 2010. Response to Klütsch and Crapon de Caprona. BMC Biology 8: 120.

Gray, M. M., Sutter, N. B., Ostrander, E. A. & Wayne, R. K. 2010. The IGF1 small dog haplotype is derived from Middle Eastern grey wolves. BMC Biology 8: 16.

Gulevich, R. G., Oskina, I. N., Shikhevich, S. G., Fedorova, E. V. & Trut, L. N. 2004. Effect of selection for behaviour on pituitary-adrenal axis and proopiomelanocortin gene expression in silver foxes *(Vulpes vulpes)*. Physiology & Behavior 82: 513-518.

Hand, J. L. 1986. Resolution of social conflicts: Dominace, egalitarianism, spheres of dominance and game theory. Quarterly Review of Biology 61: 201-220.

Hare, B. 2009. What is the Effect of Affect on Bonobo and Chimpanzee Problem Solving? In: Neurobiology of "Umwelt": How Living Beings Perceive the World, Research and Perspectives in Neuroscience ed. A. Berthoz and Y. Christen. Springer-Verlag. Berlin, Germany. pp. 89-102.

Hare, B. & Kwetuenda, S. 2010. Bonobos voluntarily share their own food with others. Current Biology 20 (5): R230.

Hare, B. & Tomasello, M. 2005. Human-like social skills in dogs?. TRENDS in Cognitive Science 9: 439-444.

Hare, B., Wobber, V. & Wrangham, R. 2012. The self-domestication hypothesis: evolution of bonobo psychology is due to selection against aggression. Animal Behaviour 83: 573-585.

Hare, B., Brown, M., Williamson, C. & Tomasello, M. 2002. The domestication of Social Cognition in Dogs. Science 298: 1634-1636.

Hare, B., Melis, A. P., Woods, V., Hastings, S. & Wrangham, R. 2007. Tolerance Allows Bonobos to Outperform Chimpanzees on a Cooperative Task. Current Biology 17: 619-623.

Hare, B., Plyusnina, I., Ignacio, N., Schepina, O., Stepika, A., Wrangham, R. & Trut, L. 2005. Social Cognitive Evolution in Capitve Foxes Is a Correlated By-Product of Experimental Domesticastion. Current Biology 15: 226-230.

Hemelrijk, C. K. 1999. An individual-orientated model of the emergency of despotic and egalitarian societies. Proceedings of the Royal Society London 266: 361-369.

Hinde, R.A. 1970. Aggression in Animals. Proceedings of the Royal Society of Medicine 63:162-163.

Jaeggi, A. V., van Noordwijk, M. A. & van Schaik, C. P. 2008. Begging for Information: Mother-Offspring Food Sharing Among Wild Bornean Orangutans. American Journal of Primatology 70: 533-541.

Kaplan, J. R., Fontenot, M. B., Berard, J. Manuck, S. B. & Mann, J. J. 1995. Delayed dispersal and elevated monoaminergic activity in free-ranging rhesus monkeys. American Journal of Primatology 35: 229-234.

Kerswell, K. J., Bennett, P., Butler, K. L. & Hemsworth P. H. 2009. The relationship of adult morphology and early social signalling of the domestic dog (*Canis familiaris*). Behavioural Processes 81: 376-382.

Kleiman, D. G. & Malcolm, J. R. 1981. The evolution of male parental investment in mannals. In: Parental Care in Mammals, ed. D. J.Gubernik & P. H. Klopfer. Plenum: New York. pp. 347-387.

Koler-Matznick, J., Brisbin, I. L. Jr. & Feinstein, M. 2005. An ethogram for the New Guinea singing dog, *Canis hallstromi*. New Guinea Singing Dog Conservation Society, Central Point, USA.

Kotrschal, K., Hemetsberger, J. & Dittami, J. 1993. Food exploitation by a winter flock of greylag geese: behavioural dynamics, competition and social status. Behavioral Ecology and Sociobiology 33: 289-295.

Kühme, W. 1965. Communal Food Distribution and Division of Labour in African Hunting Dogs. Nature 4970: 443-444.

Kummer, H. & Cords, M. 1991. Cues of ownership in long-tailed macaques, *Macaca fascicularis*. Animal Behavior 42: 529-549.

Künzl, C. & Sachser, N. 1999. The Behavioral Endocrinology of Domestication: A Comparison between the Domestic Guinea Pig (*Cavia aperea f. porcellus*) and Its Wild Ancestor, the Cavy *(Cavia aperea)*. Hormones and Behavior 35: 28-37.

Langbein, J. & Puppe, B. 2004. Analysing dominance relationships by sociometric methods – a plea for a more standardised and precise approach in farm animals. Applied Animal Behaviour Science 87: 293-315.

Leaver, S. D. A. & Reimchen, T. E. 2008. Behavioural responses of Canis familiaris to different tail lengths of a remotely-controlled life-size dog replica. Behaviour 145: 377-390.

Lindsay, S. R. 2005. Handbook of Applied Dog Behavior and Training. Volume 3: Procedures and Protocols. Blackwell Publishing Oxford, UK. pp. 535

Lockwood, R. 1995. The ethology and epidemiology of canine aggression. In: The domestic Dog: its evolution, behaviour and interactions with people, ed. J. Serpell. Cambridge University Press, UK. pp. 131-138.

Lorenz, K. 1964. Das sogenannte Böse. Zur Naturgeschichte der Aggression. 17.-20. Auflage. Dr. G. Borotha-Schoeler Verlag. Wien, Austria.

Malcolm, J. R. & Marten, K. 1982. Selection and the Communal Rearing of Pups in African Wild Dogs *(Lycaon pictus)*. Behavioral Ecology and Sociobiology 10: 1-13.

Marler, P. 1956. Studies of fighting in Chaffinches (3) proximity as a cause of aggression. The British Journal of Animal Behaviour 4: 23-24.

Matthey-Doret, S. 2010. Aggressive interactions in Canids during feeding: A comparative study of hand-raised wolves (*Canis lupus occidentalis*) and dogs (*Canis familiaris*) living in packs. Diploma thesis University of Neuchâtel, Switzerland

McLeod, P. 1990. Infanticide by female wolves. Canadian Journal of Zoology 68: 402-404.

Mech, L. D. 1970. The wolf: The ecology and behaviour of an endangered species. University of Minnesota Press, Minneapolis, USA.

Mech, L. D. 1994. Buffer zones of territories of gray wolves as regions of intraspecific strife. Journal of Mammalogy: 75:199-202.

Mech, L. D. 1999. Alpha status, dominance, and division of labor in wolf packs. Canadian Journal of Zoology 77: 1196-1203.

Mech, L. D. & Boitani L. 2003. Wolves: behavior, ecology, and conservation. University of Chicago Press, Chicago, USA.

Mech, L. D. & Adams, L., Meier, T., Burch, J. & Dale, B. 1998. The Wolves of Denali. University of Minnesota Press, Minneapolis, Minnesota, USA.

Melis, A. P., Hare, B. & Tomasello, M. 2006. Engineering cooperation in chimpanzees: tolerance constraints on cooperation. Animal Behaviour 72: 275-286.

Miklósi, Á. 2007. Dog Behaviour, Evolution, and Cognition. Oxford University Press, Oxford, UK.

Miklósi, Á, Topál, J. & Csányi, V. 2004. Comparative social cognition: what can dogs teach us? Animal Behaviour 67: 995-1004.

Miklósi, Á, Kubinyi, E., Topál, J., Gácsi, M., Virányi, Z. & Csányi, V. 2003. A Simple Reason for a Big Difference: Wolves Do Not Look Back at Humans, but Dogs Do. Current Biology 13: 763-766.

Mitani, J. C., Watts, D. P. & Amsler, S. J. 2010. Lethal intergroup aggression leads to territorial expansion in wild chimpanzees.Current Biology 20: R507-R508.

Morey, D. F. 1994. The Early Evolution of the Domestic Dog. American Scientist 82: 336-347.

Möslinger, H. 2008. Cooperative string-pulling in wolves (*Canis lupus*). Diploma thesis University of Vienna, Austria

Murray, D., Smith, D., Bangs, E., Mack, C., Oakleaf, J., Fontaine, J., Boyd, D., Jiminez, M., Niemeyer, C. & Meier, T. 2010. Death from anthropogenic causes is partially compensatory in recovering wolf populations. Biological Conservation 143: 2514-2524.

Nunn, C. L. & Deaner, R. O. 2004. Patterns of participation and free riding in territorial conflicts among ringtailed lemurs *(Lemur catta)*. Behavioral Ecology and Sociobiology 57: 50-61.

Oskina, I.N. 1996. Analysis of the functional state of the pituitary– adrenal axis during postnatal development of domesticated silver foxes *(Vulpes vulpes)*. Scientifur 20: 159– 61.

Pal, S. K. 2003. Reproductive behaviour of free-ranging rural dogs (*Canis familiaris*) in relation to mating strategy, season and litter production. Acta Theriologica 48: 271-281.

Pal, S. K. 2004. Parental care in free-ranging dogs, *Canis familiaris*. Applied Animal Behaviour Science 90: 31-47.

Pal, S. K, Ghosh, B. & Roy, S. 1998. Agonistic behaviour of free-ranging dogs (*Canis familiaris*) in relation to season, sex, and age. Applied Animal Behaviour Science 59: 331-348.

Pal, S., Ghosh, B. & Roy, S. 1999. Inter-and intra-sexual behaviour of free-ranging dogs *(Canis familiaris)*. Applied Animal Behaviour Science 62: 267-278.

Pang, J.-F., Kluetsch, C., Zou, X.-J., Zhang, A., Luo, L.-Y., Angleby, H., Ardalan, A., Ekström, C. Sköllermo, A., Lundeberg, J., Matsumura, S., Leitner, T., Zhang, Y.-P. & Savolainen, P. 2009. mtDNA Data Indicate a Single Origin for Dogs South of Yangtze River, Less Than 16,300 Years Ago, from Numerous Wolves. Molecular Biology and Evolution 26: 2849-2864.

Parker, G. A. 1974. Assessment strategy and the evolution of fighting behavior. Journal of Theoretical Biology 47: 223-243.

Peterson, R. O. & Ciucci, P. 2003. The Wolf as a Carnivore. In: Wolves: behavior, ecology, and conservation ed L. D. Mech & L. Boitani. University of Chicago Press, Chicago, USA. pp. 104-130.

Popova, N. K., Voitenko, N. N., Kulikov, A. V. & Avgustinovich, D. F. 1991. Evidence for the involvement of central serotonin in mechanism of domestication of silver foxes. Pharmacology, Biochemistry and Behavior 40: 751-756.

Preuschoft, S. & van Schaik, C. P. 2000. Dominance and communication. Conflict management in various social settings. In: Natural Conflict Resolution, ed. F. Aureli & F. B. M. de Waal. University of California Press, Berkeley, California, USA. pp. 77-105.

Robb, S. E. & Grant, J. W. A. 1998. Interaction between the spatial and temporal clumping of food afftect the intensity of aggression in Japanese medaka. Animal behaviour 56: 29-34.

Saetre, P., Lindberg, J., Leonard, J. A., Olsson, K., Pettersson, U., Ellegren, H., Bergström, T. F., Vià, C. and Jazin, E. 2004. From wild wolf to domestic dog: Gene expression changes in the brain. Molecular Brain Research 126: 198-206.

Saito, A., Izumi, A. & Nakamura, K. 2008. Food Transfer in Common Marmosets: Parents Change Their Tolerance Depending on the Age of Offspring. American Journal of Primatology 70: 999-1002.

Saito, C., Sato, S., Suzuki, S., Sugiura, H., Agetsuma, N., Takahata, Y., Sasaki, C., Takahashi, H., Tanaka, T. & Yamagiwa, J. 1998. Aggressive Intergroup Encounters in Two Populations of Japanese Macaques *(Macaca fuscata)*. Primates 39: 303-312.

Sands, J. & Creel, S. 2004. Social dominance, aggression and faecal glucocorticoid levels in a wild population of wolves, Canis lupus. Animal Behaviour 67: 387-396.

Savolainen, P., Zhang, Y., Luo, J. Lundeberg, J. & Leitner, T. 2002. Genetic Evidence for an East Asian Origin of Domestic Dogs. Science 298: 1610-1613.

Schenkel, R. 1947. Expression studies of wolves. Behaviour 1: 81-129.

Schjelderup-Ebbe, T. 1935. Social behaviour of birds. In: Handbook of Social Psychology, ed. C. Murchison. Clark University Press, Worcester, Mass. pp. 947-972.

Scott, J. & Lockard, J. S. 2006. Capitve female gorilla agonistic relationships with clumped defendable food resources. Primates 47: 199-209.

Sigg, H. & Falett, J. 1985. Experiments on the respect of possession and property in hamadryas baboons (*Papio hamadryas*). Animal Behavior 33: 978–984.

Soproni, D., Miklósi, Á,. Topál, J. & Csányi, V. 2001. Comprehension of Human Communicative Signs in Pet Dogs *(Canis familiaris)*. Journal of Comparative Psychology 115: 122-126.

Sterck, E. H. M., Watts D.P. & van Schaik, C.P. 1997. The evolution of female social relationships in nonhuman primates. Behavioral Ecology and Sociobiology 41: 291-309.

Thierry, B. 1997. Adaptation and self-organization in primate societies. Diogenes 180: 39-71.

Thierry, B. 2000. Covariation of conflict management patterns across macaque species. In: Natural Conflict Resolution, ed. F. Aureli & F. B. M. de Waal. University of California Press, Berkeley, USA. pp. 106-128.

Thierry, B. 2004. Social epigenesist. In: Macaque Societies: A Model for the Study of Social Organization, ed. B. Thierry, M. Singh & W. Kauffmanns. Cambridge University Press, Cambridge, UK. pp. 267-290.

Thierry, B. 2007. Unity in Diversity: Lessons From Macaque Societies. Evolutionary Anthropology 16: 224-238.

Toates, F. 1995. Stress: Conceptual and Biological Aspects. John Wiley & Sons Ltd., West Sussex, England. pp. 38-44.

Trut, L. N. 1999. Early Canid Domestication: The Farm-Fox Experiment. American Scientist 87: 160-169.

Trut, L. N., Plyusnina, I. Z., Kolesnikova, L. A. & Kozlova, O. N. 2000. Interhemispheral neurochemical differences in brains of silver foxes selected for behaviour and the problem of directional asymmetry. Russian Journal of Genetics 36: 776-780.

Tsigos, C. & Chrousos, G. P. 2002. Hypothalamic-pituitary-adrenal axis, neuroendocrine factors and stress. Journal of Psychosomatic Research 53: 865-871.

Ueno, A. & Matsuzawa, T. 2004. Food transfer between chimpanzee mothers and their infants. Primates 45: 231-239.

van Schaik C. P. 1989. The ecology of social relationships amongst female primates. In:. Comparative socioecology: the behavioral ecology of humans and other mammals, ed. V. Standen, R. A. Foley. Blackwell Scientific Publications: Oxford, UK. pp. 195–218.

van Schaik, C. P. & van Noordwijk, M. A. 1988. Scramble and Contest in Feeding Competition among Female Long-Tailed Macaques *(Macaca fascicularis)*. Behaviour 105: 77-98.

Vogel, E. R., Munch, S. B. & Janson, C. H. 2007. Understanding escalated aggression over food resources in white-faced capuchin monkeys. Animal Behaviour 74: 71-80.

vonHoldt, B. M., Pollinger, J. P., Lohmueller, K. E., Han, E., Parker, H. G., Quignon, P., Degenhardt, J. D., Boyko, A. R., Earl, D. A., Auton, A., Reynolds, A., Bryc, K., Brisbin, A., Knowles, J. C., Mosher, D. S., Spady, T. C., Elkahloun, A., Geffen, E., Pilot, M., Jedrzejewski, W., Greco, C., Randi, E., Bannasch, D., Wilton, A., Shearman, J., Musiani, M., Cargill, M., Jones, P. G., Qian, Z., Huang, W., Ding, Z.-L., Zhang, Y.-p., Bustamante, C. D., Ostrader, E. A., Novembre, J. T. & Wayne, R. K. 2010. Genome-wide SNP and haplotype analyses reveal a rich history underlying dog domestication. Nature 464: 898-903.

Wilson, M. L. & Wrangham, R. W. 2003. Intergroup Relations in Chimpanzees. Annual Review of Anthropology 32: 363-392.

Wrangham, R. W., Wilson, M. L. & Muller, M. N. 2006. Comparative rates of violence in chimpanzees and humans. Primates 47: 14-26.

Zimen, E. 1990. Der Wolf. Verhalten, Ökologie und Mythos. Knesebeck & Schuler GmbH & Co. Verlags KG, München, Germany.

Zimen, E. 1992. Der Hund. Abstammung – Verhalten – Mensch und Hund. Wilhelm Goldmann Verlag, München, Germany.

6 Appendix

6.1 Ethogram

<u>Aggressive Behaviours</u>

- **Weak aggression**

Ambush	lying in a sphinx posture, staring intently at another wolf or at prey
Bark	first a breathy woof, soft; second the growl bark with growling overtones; third a bow wow is a bark out loud without any growly overtones.
Besnuffle	to sniff in an exaggerated way, tends to precede bites.
Chase	running in pursuit of another one
Dominant approach	to go forward within 2 m to another individual with the tail perpendicularly or above the plane of the back and the ears erects and pointed forward and head held high
Growl	a throaty rumbling vocalization, usually low in pitch
Guard	to stay by something and to drive others away from it
Stalk	to pursue another individual or prey by means of stealthy approach; the head level is lower than the top of the back with the ears directed forward

- **Intermediate aggression**

Displace	aggressor causes opponent to move away from a resource or goal
Grab	to bite an object or another individual, and hold on firmly
Inhibited bite	a bite without sufficient pressure to wound a canine
Muzzle bite	grabbing the muzzle and applying enough pressure to make the grabbed individual whimper
Pin	to grab another individual forcing it to the ground and holding it there
Pull	to grab another individual and draw it along, without the pulled individual being recumbent
Push away	to use feet and legs both defensively while engage in a ritualized aggression

Rebuff	rejecting a suitor and driving him or her away
Ride up	to mount another one from behind or from side
Snap	a rapid bite that has a little contact with its object; as the canid's jaws come together, the teeth make an audible sound
Tug of War	two individuals taking hold of different parts of an object and tugging vigorously against each other

- **Strong aggression**

Attack	a running or jumping approach toward another one, often bites at the neck or muzzle forcing it on the ground and holding it there
Bite	to close jaws and teeth hard on someone
Fight	high intensity, aggressive, often damaging encounters
Hipslam	the individual pivots on its forepaws and slams into its target with its hindquarters
Jaw spar	two canids "fencing" with open jaws; as they block each other's feints, neither actually closes its jaws
Knock down	striking another individual with a sharp blow, usually with the chest and shoulders
Mob	chasing, jaw sparring, biting, and/or wrestling or pinning by two or more individuals orienting to a third

Submissive Behaviours

Active submission	the individual has its tail tucked between his hind legs, sometimes wag it while he is in a crouched position and may attempt to paw and lick the side of aggressor's muzzle and mostly pees
Flee	to walk or run with tail tucked and body ducked away from other individuals
Food beg	to lick, nibble, pull or paw at another individual's muzzle and lips
Crouch	to lower the head, bent the legs, the back often arched and the tail between the legs. The wolf looks hunched and smaller

| Passive submit | to lie on the back, demonstrate the stomach and the tail is between the legs; the are ears directed backwards and close to the head and raises a hind leg for inguinal presentation |
| Inguinal present | to show the groin area to another subject often while laying on the back |

Dominant Behaviours

Dominant approach	to go forward within 2 m to another individual with the tail perpendicularly or above the plane of the back and the ears erects and pointed forward and head held high
Genital sniffing	to sniff the genital parts of another individual when this one is lying on the back in the passive submission position
Mark	to urinate with the hind leg lifted up in the air mostly near or on bushes or on a tree
Muzzle bite	grabbing the muzzle of another individual
Stand over	to stand over opponent's body, or place the forepaws on the opponent and over the negative response from the opponent which is growling and trying to get out of the situation
Stand tall	an individual draws itself up to its full height; the neck is often arched and the ears pricked
T-position	one individual approaches the shoulder region of another one and often puts its head on its shoulder; formation looks like the capital "T"
Ride up	to mount another one from behind or from side

7 Acknowledgment

First of all I want to thank Prof. Dr. Kurt Kotrschal to be my supervisor and Dr. Zsofi Virányi and Dr. Friederike Range to give me the opportunity to collect my data at the WSC and to help me with the thesis and give me advise.

I also want to thank the trainers at the WSC for helping me with my experiments and giving me inside in the training of the wolves and dogs. You showed me the way which I finally want to go in my life.

A special thank goes to Marianne Heberlein who helped me so much with the feedings and especially with the statistics, without you I couldn't write this now and for becoming such a good friend. I also want to thank Ewelina Utrata, Marion Heszle, Caroline Ritter and Patricia Berner for helping me whenever I needed something and for the amazing social support. You really made this 7 month special and unique.

The biggest thank you of all is for the animals, the wolves and dogs. Thank you for letting me in your world for accepting me and for giving me such a nice time. You are incredible and I will never forget you, especially Cherokee, Wapi and Kali who had to leave this world far too early. You will be in my heart forever.

Last but not least I would like to thank my parents for the support and the possibility to concentrate on my study.

Printed by Books on Demand GmbH, Norderstedt / Germany